The Natural Method

The Natural Method

Essays on Mind, Ethics, and Self in Honor of Owen Flanagan

Edited by Eddy Nahmias, Thomas W. Polger, and Wenqing Zhao

The MIT Press
Cambridge, Massachusetts
London, England

This book was set in Stone by Westchester Publishing Services, Danbury, CT. Printed and bound in the United States of America.

Library of Congress Cataloging-in-Publication Data

Names: Nahmias, Eddy A., 1969- editor. | Polger, Thomas W., editor. | Zhao, Wenqing, 1989- editor. | Flanagan, Owen J., honoree.
Title: The natural method : essays on mind, ethics, and self in honor of Owen Flanagan / edited by Eddy Nahmias, Thomas W. Polger, and Wenqing Zhao.
Description: Cambridge, Massachusetts : The MIT Press, [2019] | Includes bibliographical references and index.
Identifiers: LCCN 2019035888 | ISBN 9780262043991 (hardcover)
Subjects: LCSH: Flanagan, Owen J.--Influence. | Philosophy of mind. | Consciousness. | Ethics.
Classification: LCC BD418.3 .N345 2019 | DDC 191--dc23
LC record available at https://lccn.loc.gov/2019035888

10 9 8 7 6 5 4 3 2 1

To our mentor, Owen

Contents

Foreword

Patricia S. Churchland

A celebration in honor of Owen Flanagan and his storied career in philosophy stirs up memories of his tenacious curiosity about all things, along with his willingness to size up problems without being hemmed in by philosophical fashions.[1] It is also an occasion to take a long look at the academic discipline of philosophy over the last forty years or so and its sometimes quite strict orthodoxies. The glance backward is especially fitting when the connecting thread is Flanagan and the way he went about things or, perhaps more accurately, did *not* go about things.

Especially in the second half of the twentieth century, many philosophers tended to be deferential to the "mainstream" and thus to the luminaries who set the philosophical agenda, but not Flanagan. He could be respectful, but he did not shape his inquiries to fit trendy theories and methods. For example, when others were pushing "necessary truths" rooted in their convictions about what words *do* mean and *must* mean, Flanagan preferred to take a good look at what the data revealed about the phenomena the words described. When others were assembling natty metaphors to "explain" consciousness, Flanagan was the head-scratching skeptic asking what explanatory value the metaphors cornered and how they measured up against scientific data. When moral philosophers lined up to proclaim themselves either Kantians or Utilitarians and do battle accordingly, Flanagan wisely saw that there was a far deeper story that both factions totally missed.

Fads and fashions seem to find a welcome home in philosophy, which is why the Cambridge group, Beyond the Fringe, and their later Oxfordian counterpart, Monty Python's Flying Circus, found such ready material. The satire staged in 1977 by John Cleese and Jonathan Miller of a common room conversation between two philosophers concerning what philosophers do remains surprisingly current. At one point, Cleese responds to some banal point of Miller's by saying "yes, yes" and Miller sincerely queries, "Do you mean 'yes' in the affirmative sense,'" to which Cleese, after a thoughtful pause, says, "no."[2] The entire caricature is stunningly silly, and yet the echoes of actual philosophical practice are all too audible. Of course, fads and foolishness can be found in science too, but perhaps any discipline is especially vulnerable to fads when little is known concerning its chosen topics, thereby rewarding debating swagger and sometimes plain silliness. For example, this was true concerning the nature of the heart until William Harvey discovered in the early seventeenth century that it was a meat pump.

What Flanagan realized early on, and what motivated his first book, *The Science of the Mind*,[3] is that if we want to understand phenomena such as consciousness, decision-making, and reasoning, we will not get very far unless we have evidence, as opposed to mere speculations, metaphors, and dogmatic edicts, about what certain words mean. Thus, he felt at home with naturalism, the very general view that understanding natural phenomena requires, among other things, paying attention to the facts regarding the phenomena. Or, to paraphrase Quine in his defense of naturalism, there is no principled division between science and philosophy.

Although Flanagan's heresy on how to make progress might have been a break with trends in philosophy circa 1980, it was not a break with the grand philosophical tradition of Aristotle, René Descartes, David Hume, Adam Smith, William James, and Hermann von Helmholtz. Rather, it was an extension of that same tradition. At this point in the twenty-first century, Flanagan's unfashionable regard for the empirical world of experiments, data, and how science makes progress has paid off—handsomely. By contrast, some of the hotly contested ground of the 1990s now looks a bit forlorn, more abandoned than productively cultivated.

As Flanagan well appreciated, neuroscience, ethology, anthropology, and experimental psychology have made discoveries that, if we permit, have an impact on traditional philosophical questions, such as the nature of decision-making, consciousness, and the origins of moral values.

As we look to the future, the nature of knowledge, learning, and memory is a broad topic where the impact of neuroscience on philosophical theories is likely to be substantial, if not outright revolutionary. Any epistemologist worth her salt will be intrigued by the discovery that nonverbal animals clearly have spatial knowledge, such as where home is and where the food is, as well as nonspatial knowledge, such as who one's mate is. Like us, nonverbal animals learn from experience, and what is learned is put to good use in their behavior. Neuroscientists have inquired into the mechanisms that support this knowledge, and impressive data relevant to these mechanisms have been discovered. In 2014, the Nobel Prize in Physiology or Medicine went to Edvard Moser, May-Britt Moser, and John O'Keefe "for their discovery of cells that constitute a positioning system in the brain."

It turns out that the hippocampus and connected structures (e.g., medial entorhinal cortex, pre-subiculum) are essential for spatial knowledge as well as for episodic memory generally (e.g., what music I listened to at breakfast). There is a demonstrable link between repeatable patterns of activity of the neurons of the hippocampus of a rat and the spatial routes it uses to navigate the world it inhabits. In short, the hippocampus and ancillary structures construct organized, flexible, and usable representations of the rat's spatial environment. Spatial knowledge is only part of what the hippocampus constructs. Other cognitive functions, such as social knowledge of who to trust, also depend on the hippocampus.

The neural mechanisms supporting spatial cognition are complex, though major clues derive from the precise tuning properties of distinct groups of neurons. Some neurons (place cells) respond preferentially whenever the rat enters a particular place in an area such as a room. Other neurons (head direction cells) respond specifically to the direction in which the rat's head is pointing. Yet other neurons respond to a specific distance from the *borders* of the region that the rat is in. Remarkably, in the *medial entorhinal cortex*, which feeds into the hippocampus, a population of neurons collectively tiles the entire space that the rat is in. These neurons are called *grid cells*, and they allocentrically represent a spatial area (as opposed to merely a response to sensory cues in the area). When the rat explores a new room, these neurons then tile that new area. Under those conditions, the hippocampal place cells remap as well. Moreover, the upshot of this complex spatial framework is that the rat knows *where it is* in the larger environment. It knows where to go to get what it seeks. The rat need not see the food or the hiding place to make a beeline for it.[4]

Exactly how the different types of cellular signals are integrated to yield the systematic representations of space and other forms of relational representations (such as remembering the flow of events across time and the ability to draw novel inferences from memory) are not yet well understood.

The rat's spatial knowledge is more abstract than one might have guessed, as is illustrated by recording the activity of neurons. When a rat is resting or merely planning to forage, an observable pattern of activity is seen in the rat hippocampus that replicates the activity seen, for example, when it is actually running that successful route to the food cache. In short, its brain rehearses the winning route even while not running the route, and the time taken is compressed. This is probably a lot like what I do when I rehearse (i.e., imagine) a good route to a chanterelle mushroom patch in the woods. I would say, and many neuroscientists do say, that the rat knows where the food is likely to be found much as I know my route to the chanterelles. When a rat confronts a range of possible routes, its brain will play out (i.e., imagine) a series of routes before making a selection. This is probably a lot like what I do when I confront a range of routes for a portage of my canoe, some of them more hazardous than others. This rehearsal of routes involves representing the future—where I could go or will go—not just the past (what I did last time). In short, there is a lot of smart cognitive business going on here. This is not simple stimulus-response conditioning.

For philosophers to appreciate the epistemological significance of hippocampal discoveries made over the last four decades, they will find it useful to know as much as possible about the physiology of the hippocampus and its role in behaviors such as foraging and escaping predators. Along the way, they may bump into developmental psychologists, such as Jean Mandler, who are testing the hypothesis that spatial primitives (such as "close by" and "far away") in baby humans are the scaffolding for many categories that babies come to acquire as they develop.[5] They will perhaps see a link to the vast array of spatial metaphors common in so many languages that scaffold the way we make inferences.[6] Undoubtedly they will be more productive if they relax the rigid dogma that "knowing how" involves a completely different system from "knowing that," along with the favored assumption that believing something always requires having a language in which to represent what is believed.

One dividend of absorbing the hippocampal research is likely to be progress on intentionality, otherwise known as the *aboutness* of representations. Put another way, the question is how representations can represent something in the world, including something in the body and brain itself, such as metacognitive representation, where a monkey or a human knows what it does know. Although intentionality is sometimes considered an exclusively philosophical topic, at this stage the topic appears to be sorely in need of fresh ideas. Consequently, it may be productive to reflect on the neurobiological discoveries showing that representations in neural networks are about spatial and temporal relationships in the world. The allocentric system of grid cells is undoubtedly about some small part of the world. As the mechanisms sustaining a "positioning system" come to be more fully understood, will that help explain what it is for activity in neurons to be about things in the world? A good bet says *yes*.[7]

Should the philosophical contribution consist in divining the deep conceptual truths that limit what the neuroscientists can meaningfully say or discover about the nature of representation—of knowledge and belief? This is not likely to be a rewarding strategy. Frequently in science, we have to make do with an imprecise expression to refer to a phenomenon, at least until the science develops to the point where greater precision is possible. *Gene* is a well-known case in point. To make that term more precise, scientists in the 1940s did not turn to philosophers to reveal the conceptual truths about the concept "gene." They slowly got more data, whereupon the meaning of the word *gene* became better defined in tandem. Many expressions used in neuroscience are like *gene* circa 1940, because much about how the brain works remains poorly understood. The words *represents* and *believes* are two examples, and their meaning is apt to change profoundly over the coming decades as more is understood about how the brain learns, knows, and represents. Although immersing yourself in the neurobiological data is no guarantee of a theoretical breakthrough, it pretty obviously seems a necessary condition.

Although Flanagan was not especially focused on *represents*, he had long suspected that a lot of heavy philosophical posturing about what words must mean (supposed conceptual truths) did not accurately reflect what was going on in the actual world of language, science, and practical reasoning. All too often, what are shopped as conceptual truths anchored by word meanings are actually empirical speculations floating free of an empirical base.

As I look on the Flanagan shelf in my library, I tend to think my favorite Flanagan book is *The Bodhisattva's Brain: Buddhism Naturalized*. It taught me a lot about Buddhism and about philosophy for the ages, but Flanagan's audacity also made me laugh a lot. The first two paragraphs are quite astonishing, and bear repeating:

> Suppose we permitted ourselves this luxury: invite Confucius, Siddhartha Gautama, Mohammed, Joan of Arc, Catherine the Great, Karl Marx, Thomas Jefferson, Sojourner Truth or any other interesting or wise dead person with a view, who is representative of a tradition, into our conversations about our problems—poverty, health care, capitalism, how to be a good person, how to live well, to flourish, to be happy, and listen to what they say. This is anachronistic. Some say anachronism is bad, even that it is not allowed. Allow it.
>
> Next imagine responding to the anachronistic answers of our respected ancestors with our own reflective standards of cogency, wisdom, and breadth and depth, feeling free to judge their answers as helpful or inadequate for our problems in our time. This is ethnocentric. It is temporally different but logically identical to judging the ideas and ways of other contemporary peoples as well suited for us or not suited for us or not good for us in our time. Some say ethnocentrism is bad, even that it is not allowed. Allow it.

Thus, Flanagan launched a deep, thoughtful, and always fresh exploration of an ancient and still pressing question: what is it to live well? Flanagan's approach never appeals to a divine authority, to the edicts of pure reason, or to some shattered algorithm that aims to solve all moral problems. He knows all those pitfalls inside out. Such knowledge motivates his advice that when we are stuck on a moral problem, we need to go into the world; there is nowhere else to go. In *The Bodhisattva's Brain*, into the world is indeed where he takes us. He realizes perfectly well that what he is doing is not "the way things are done in philosophy." Instead, Flanagan is inviting us to give our curiosity space to stretch, take in oxygen, see with new eyes, hear with new ears—and to think for ourselves. Rather like Socrates, he shows by example that we should admit our ignorance, avoid arrogance and self-deception, and all the while resist getting blinded by what is fashionable. This is philosophy worth celebrating.

November 21, 2018

Notes

1. Particular thanks are owed to John Kubie, who taught me a lot about the hippocampus, and to John Kubie and Eddy Nahmias, who helped correct an earlier draft.

2. https://www.youtube.com/watch?v=qUvf3fOmTTk.

3. Owen Flanagan, *The Science of the Mind* (Cambridge, MA: MIT Press, 1984; repr. 1991).

4. For a much more complete review article, see Neil Burgess, "The 2014 Nobel Prize in Physiology or Medicine: A Spatial Model for Cognitive Neuroscience," *Neuron* 84, no. 6 (2014): 1120–1125.

5. Jean Mandler, "The Spatial Foundations of the Conceptual System," *Language and Cognition* 2, no. 1 (March 2010): 21–44.

6. George Lakoff, *Women, Fire, and Dangerous Things* (Chicago: University of Chicago Press, 1987).

7. Incidentally, the *Stanford Encyclopedia of Philosophy,* typically very up-to-date, does not yet have an entry for spatial knowledge that even mentions the hippocampus, and no general entry for spatial knowledge at all. This omission suggests a bonanza for an epistemologist with pluck and vision.

Introduction

Eddy Nahmias, Thomas W. Polger, and Wenqing Zhao

It is our pleasure to present this book in honor of our friend and mentor, Owen Flanagan. The essays collected herein span the range of his interests across his career to date. Some discuss his work directly; others are inspired by his work or his methodology. Like Owen's own work, these essays are variously penetrating and synoptic, cautious and speculative. We are very grateful to the contributors to this collection, as well as to the participants at a conference honoring Owen at Duke University in autumn 2017, and to Philip Laughlin and the staff at the MIT Press who brought this volume into being. We would also like to thank Owen's family, who welcomed us in our time as his students and colleagues and allowed him to share his time with us. Finally, and especially, we thank Owen, who always treated us as fellow explorers on a philosophical journey, never as mere passengers.

This is not the place for a full biography or bibliography of Owen Flanagan. He has yet to write his memoir as such, but his philosophy is full of autobiography. In short, Owen Flanagan was born and raised in Westchester County, New York, the oldest child of six raised in the American Irish Catholic tradition. He attended Fordham University, where he earned his undergraduate degree in 1970, and then Boston University, where he earned his PhD in 1978 under the supervision of Judson Webb. That year he took a professorship at Wellesley College, where he remained until 1993, when he moved to Duke University. He is the James B. Duke Professor of Philosophy and co-director of the Center for Comparative Philosophy. In his first forty years as a professor he has published ten books, co-edited six volumes, written over 100 articles and chapters, and mentored dozens of students. Owen's work is a model for how to be a naturalistic and scientifically informed philosopher and for how to write beautifully and deeply about a diverse range of topics, from consciousness to Buddhism, from moral psychology to dreaming, from identity to addiction, from literature to neuroscience. He writes about these topics in an eminently readable and engaging style. At the same time, it is hard for anybody to fail to notice the originality and depth of the arguments and positions he develops. Often, Owen puts in prose what we readers then realize we already believed in a way that helps us to recognize *why* we believe them.

It may seem peculiar to consider what some would call juvenilia, but Owen's dissertation already has the hallmarks—perhaps we should say "sprouts"—of his mature philosophical thinking. He posed the question, as he put it: "Does B. F. Skinner's radical behaviorism constitute a potentially adequate science of human behavior?" Owen was already concerned that the dominant psychological science of the day might fail. He had two main worries. First, a science of the mind might fail by improperly restricting its subject matter to only a subset of the behaviors in which human beings engage. Second, it might fail by not having the resources to make all the distinctions among human behaviors that are central to moral theories, for instance, between purposeful and non-purposeful behaviors. These dual concerns, individually and in combination, permeate Owen's work.

The first big concern from Owen's dissertation that became a theme in his subsequent work is that the sciences of the mind and philosophy will be individually and collectively impoverished if they artificially restrict their attention to only some kinds of evidence or reasoning. As the sciences developed through the latter part of the twentieth century—especially the cognitive and brain sciences—so did Owen's concerns. His worry that radical behaviorism neglects some aspects of human psychological behavior evolved into the concern that cognitive psychology may neglect information both from neurosciences and from persons' own reports of their experiences, particularly with respect to conscious experience or their phenomenology, and then into a concern about the neurosciences' tendency to exclude phenomena not amenable to their instruments. At the same time, Owen came to be worried that the contemporary Anglo-American conception of the scope of psychological and brain sciences were as a whole neglectful of the range of human experiences, activities, and ways of living that had been more fully appreciated not only in other cultures but even by turn-of-the-twentieth-century "Western" thinkers like William James, one of his heroes. As Owen illustrated in *The Science of the Mind* (1984), these thinkers were "always already" interdisciplinary because their theorizing was not yet constrained by the boundaries and presuppositions that came to typify institutionalized academic psychology and philosophy as the century wore on. Owen pointed out that the very narrowness of focus and specialization of knowledge that enables the development of expertise in contemporary sciences and philosophy also creates blinders that prevent us from seeing the big picture to which all specialists take themselves to be contributing. This concern applies equally to behaviorism, to the cognitive sciences, to the 1990s explosion of psychopharmacology and cognitive neuroscience, and to molecular neurobiology.

This very concern arises also for "Western" thinkers in science and philosophy writ large, and therefore Owen's views evolved from the worry that Anglo-American philosophers and scientists should consider as data the different ways of living and being that are exemplified by different cultures, to the even broader view that Anglo-American science and philosophy will be hindered if they fail to engage not only with other cultures but also with the philosophy of those cultures. That is, we should not exclude other peoples and culture as data, but we must furthermore engage and experiment with their ways of thinking and theorizing, or more generally the ways of engaging in inquiry that are practiced by those peoples and cultures.

In this way, Owen's enthusiasm for comparative and cross-cultural philosophy, especially Asian philosophy, is a natural evolution of his concern, going all the way back to his dissertation, about impoverishing our scientific and philosophical worldviews by unduly restricting the phenomena that we consider before we even begin. His concerns also include the specific experiences of individuals, which are often neglected by systematic science, be they the experiences of dreamers, addicts, patients, or bodhisattvas. Owen first codified his inclusive methodology in his illustration of the "natural method" in *Consciousness Reconsidered* (1992), where he was specifically engaged with the alternative neo-behaviorist view recommended by Daniel Dennett's "heterophenomenology."[1] Owen continued to elaborate that picture, literally and metaphorically, in *Self-Expressions* (1996), *Dreaming Souls* (1999), and other works. His philosophical method is "natural" in several ways. It is natural in the Quinean sense of taking philosophy and the sciences to be continuous, none being privileged over the others *a priori*. It is natural in the empiricist sense of relying on observations to drive theories, though it is much more expansive than that of most empiricists about what may count as an observation. And, like Penelope Maddy's "Second Philosopher," Owen finds it obvious and unobjectionable that we may make use of whatever epistemic resources we in fact find available to us, so that we are "natural born" naturalists.[2]

The second big concern in Owen's dissertation that became a theme of his work is that scientific theories might not be adequate for the philosophical work for which they are needed. In his dissertation he took for granted that it would be a failing of radical behaviorist psychology as a science if it could not make the distinction between purposeful and nonpurposeful behaviors. (He argued that it could.) But the concern is quite general: philosophical theories frequently presuppose claims about how the world is. Many of these claims can be checked by current sciences, and where there is disagreement, the science and philosophy should be brought into some kind of broad reflective equilibrium, as required by Owen's natural method. A special case of this concern is that moral claims and ethical theories invariably make assumptions about human nature; that is, about

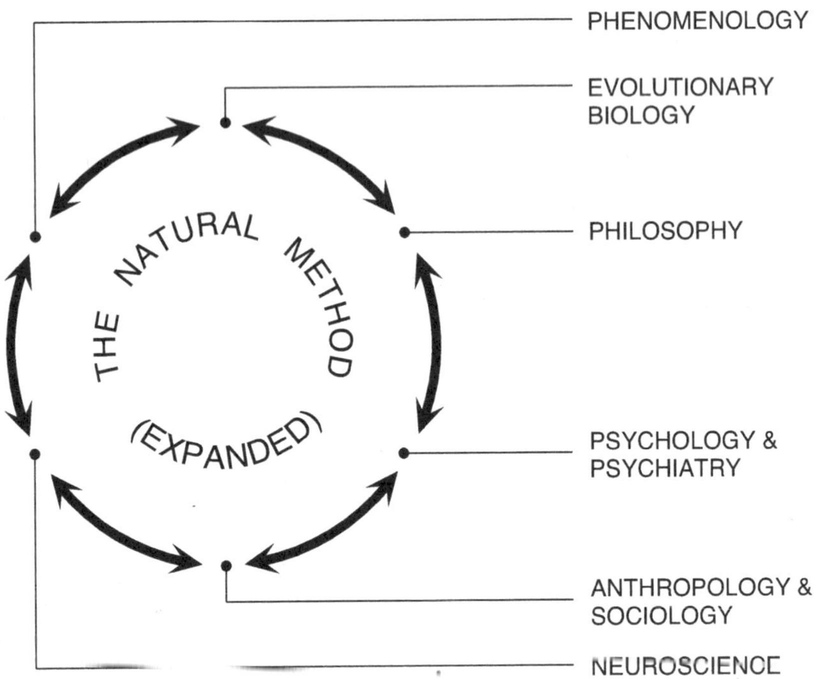

Flanagan's natural method.

persons. In particular, they make assumptions about the motives and capacities of humans as moral agents and subjects—about moral psychology, a now active interdisciplinary field that Owen's work helped to revitalize. In *Varieties of Moral Personality* (1991), Owen argued that facts about human psychology ought to constrain moral theorizing in accord with what he called the Principle of Minimal Psychological Realism: "Make sure when constructing a moral theory or projecting a moral ideal that the character, decision processing, and behavior prescribed are possible, or perceived to be possible, for creatures like us."[3]

Owen always recognized that this principle leaves open many questions. For example, about which creatures are relevantly "like us" and what sort of "possibility" is at stake. Originally, he had in mind questions such as whether human beings in fact have the kinds of stable character traits postulated by Aristotelian ethical theories, but the form of the question is familiar to anyone who has doubted that human beings have the kinds of "archangelic"[4] reasoning powers required by many consequentialist moral theories or that we should aim to be "moral saints."[5] Although in *The Geography of Morals*[6] Owen subsequently backed off stronger versions of the Principle of Minimal Psychological Realism, he has not abandoned his general view that philosophy and the sciences must be brought into broad reflective equilibrium. Indeed, Owen's reasons for softening the Principle of Minimal Psychological Realism come from, as it were, taking his own medicine: he came to recognize that the moral theories of diverse cultures (and perhaps even those of Western monotheism) do not in fact invariably accept the metaethical constraint that "ought implies can" and therefore do not require what they prescribe to be possible for creatures like us. This transition in his thinking was prompted in part by his participation in the Dalai Lama's Mind and Life conference in Dharamasla, India, in 2000. The topic of the meeting was destructive emotions, and the experience there proved to be formative for Owen.

These two general themes of Owen's philosophical work are familiar to his students, colleagues, and readers. Those who have come to know him in person are also familiar with his good-humored and often sympathetic approach to philosophy and philosophical conversation—neither altogether tender-minded nor tough-minded, in William James's terms. The recognition that philosophy bumps up against the sciences manifests, in us his students, in a deep skepticism of either philosophical or scientific assertions about how "we" are or must be. And the commitment to finding some way to unify philosophical and scientific viewpoints requires a commitment to interdisciplinary inquiry and optimistic temperament, combined with a willingness to change one's mind.

In this manner, Owen embodies—and his approach recommends—a sense of intellectual humility that is sometimes missing in contemporary philosophy. We hope that the essays in this book will inspire others to follow his example, even or especially if they disagree with some of his particular conclusions.

Atlanta, Georgia
Cincinnati, Ohio
Walla Walla, Washington
February 2019

Notes

1. Daniel Dennett, *Consciousness Explained* (Boston: Little, Brown, 1991).

2. Penelope Maddy, *Second Philosophy* (New York: Oxford University Press, 2007).

3. Owen Flanagan, *Varieties of Moral Personality: Ethics and Psychological Realism* (Cambridge, MA: Harvard University Press, 1991), 32.

4. R. M. Hare, *Moral Thinking: Its Levels, Method and Point* (Oxford: Clarendon Press, 1981).

5. Susan Wolf, "Moral Saints," *Journal of Philosophy* 79, no. 8 (August 1982): 419–439.

6. Owen Flanagan, *The Geography of Morals: Varieties of Moral Possibility* (New York: Oxford University Press, 2016).

Monographs by Owen Flanagan

The Science of the Mind (MIT Press, 1984; 2nd ed. 1991).

Varieties of Moral Personality: Ethics and Psychological Realism (Harvard University Press, 1991).

Consciousness Reconsidered (MIT Press, 1992).

Self-Expressions: Mind, Morals and the Meaning of Life (Oxford University Press, 1996).

Dreaming Souls (Oxford University Press, 1999).

The Problem of the Soul: Two Visions of Mind and How to Reconcile Them (Basic Books, 2002).

The Really Hard Problem: Meaning in a Material World (MIT Press, 2007).

The Bodhisattva's Brain: Buddhism Naturalized (MIT Press, 2011).

Moral Sprouts and Natural Teleology: 21st Century Moral Psychology Meets Classical Chinese Philosophy, the 78th Aquinas Lecture (Marquette University Press, 2014).

The Geography of Morals: Varieties of Moral Possibility (Oxford University Press, 2016).

1 *Consciousness Reconsidered* Reconsidered

Robert Van Gulick

For a long period running at least from René Descartes in the early seventeenth century[1] to William James at the end of the nineteenth,[2] consciousness was central to our conception of mind. Yet, consciousness largely disappeared from psychological and scientific theorizing about the mind for most of the twentieth century. In the United States, that resulted partly from the rise of behaviorism, which denied the importance not only of consciousness but also of mentalistic theories in general. However, even with the so-called cognitive revolution at midcentury and the return of robust theorizing about internal mental processes, consciousness remained largely ignored until almost the close of the century. Not until the 1980s did consciousness return as a subject of serious scientific investigation and significant philosophical discussion, but in the past few decades there has been explosive growth in research on consciousness and a wealth of empirical and philosophical theories about its features, basis, and its place in the natural world. The problem of consciousness thus presents an excellent case study for assessing the natural method in action.

Owen Flanagan was in the vanguard of the contemporary movement to investigate consciousness. Just over a quarter century ago, in 1992, when the consciousness revolution was in its very early stages, Flanagan published *Consciousness Reconsidered* (1992), a work that grew out of a seminar that he had taught at Wellesley College at which a number of other philosophers working on consciousness had been visiting speakers, myself included. *Consciousness Reconsidered* provides a snapshot of that historical period, both of the prospects for a scientific and philosophical understanding of consciousness as well as some of the critical challenges raised at that time by those who were skeptical about our ability to explain consciousness in a naturalistic and intelligible way. Thus, this volume honoring Flanagan's many important contributions offers the perfect occasion to assess to what degree the naturalistic project of understanding consciousness that he proposed succeeded in the past two and a half decades.

At the outset of *Consciousness Reconsidered,* Flanagan observed that there are three great mysteries that have puzzled humanity for much of its history: Why is there something rather than nothing? What is life? What is consciousness? This leads to two further questions: What is the status of those three great enigmas today in the second decade of the twenty-first century? And, given our current understanding, which, if any, of them has been solved, and which remain unanswered?

The question of why anything at all exists may be the most basic and elemental of the three and might be equated with the cosmological question of the origin of the universe: "What is the origin of the entirety of the world of space, time, energy, and matter?" Historically, most answers have appealed to a creator or god, a supernatural being powerful enough to bring an entire universe into existence from nonexistence. Contemporary physics gives a very different sort of explanation, one that appeals to the so-called big bang, which marked the origin of the universe approximately 13.8 billion years ago in an explosive expansion of time, space, and energy. Using available evidence and theory, physicists have been able to retrodictively infer the basic state of the universe back to just nanoseconds after the big bang and to spell out the general stages and path by which the universe evolved to its present state. Moreover, plausible models have been offered to explain the origin of the big bang itself. In particular, so-called cosmic inflation models explain how a rapid expansion of space-time outward from a singularity could result from random quantum fluctuations in the zero-energy background state. Because the background state is one of zero energy (i.e., a state of no energy), there is a sense in which cosmic inflation models may seem to violate a basic metaphysical principle that has been espoused since antiquity: "Ex nihilo, nihil fit," or "From nothing, comes nothing." However, that may depend on how one interprets the zero-energy state. Is it a *nothing* because it has zero energy? Or is it still metaphysically a *something*—though admittedly something strangely quantum mechanical and not within our universe or our space-time? If one opts for the latter, then a physical explanation of why our universe exists would not suffice to answer the deep metaphysical mystery of why there is something rather than nothing. The problem would recur at the prior level. Thus, the status of the first mystery remains in dispute. Quantum cosmology may be able to explain the origin of the universe, but it is less clear that it suffices to explain why there is anything at all.

The status of the second mystery changed radically in the middle of the last century. In the early twentieth century, there was still considerable scientific disagreement about the nature of life and about the possibility of explaining living organisms in terms of physical processes or mechanisms alone. The mechanist view had many supporters, but others rejected it. Vitalists appealed to extra vital forces unique to living things in order to explain biological processes. Organicists rejected vitalism and its appeal to special forces but nonetheless argued that living organisms could not be explained or understood solely in terms of physical mechanisms and processes. The organicist view was strongly antireductionist. For example, noted British chemist and physiologist J. S. Haldane wrote, "Even by the widest stretch of the imagination it is impossible to conceive of any mechanism capable of bringing about the phenomena, fundamentally characteristic for life, of organic maintenance and reproduction."[3]

The discovery in 1954 of the structure of DNA, and soon afterward of its basic genetic coding system, ushered in the modern era of molecular biology and genetics, with its explosive growth in our understanding of the detailed mechanisms of heredity and how they control, construct, and regulate the full array of subcellular structures and processes. The models continue to develop, and new molecular mechanisms come to light, such as the recognition in recent decades of the importance of epigenetic factors in activating and silencing genes. Further additions are sure to be needed, and there is admittedly a great deal we do not yet understand about all the particular details in any specific organism. We cannot trace every step that leads from a fertilized moose zygote to an adult moose. Nonetheless, we are confident that we know the basic story of how "the trick is done" with respect to life.

With respect to the initial origin of life, there is as yet no generally accepted scientific explanation, but there are numerous plausible hypotheses about the physical contexts in which molecules known to have existed in the prebiotic environment might have energetically combined to produce self-sustaining, self-reproducing, membrane-enclosed systems, which could have evolved into systems that are genuinely alive. What the actual pathway was remains unclear, and new options have recently been proposed, including those inspired by the discovery of rich biological ecosystems surrounding deep-sea vents that depend on geothermal and chemical energy rather than sunlight and photosynthesis as the basis of their food and energy chain. So, while the origin of life remains an unsolved scientific problem, there is ample reason to be optimistic about our long-term success and to believe that when we do have an answer it will be generally consistent with our current basic biomolecular understanding of the nature of life. The air of mystery has been dispelled.

What then of the third mystery, that of consciousness? It shares a form common to the other mysteries. Each of the three is a variant of the basic question type "How do you get something that is F from things that are not F?": How do you get something that *exists* from what does *not exist*? How do you get things that are *alive* from things that are *not alive*? How do you get something that is *conscious* from things that are *not conscious*? Despite their similarity, the consciousness mystery is the most specific of the three, and success answering the first two need not guarantee success answering the third. Even if we were to understand why there is something (a universe) and how living things exist, we still might be puzzled about how any of those things—living or not—is conscious. It seems we can imagine a possible world with physics, chemistry, and even biology, but nothing that is conscious. Atoms, mountains, oceans, and even living things might exist in the total absence of consciousness, or so at least it seems—panpsychists or idealists to the contrary.

Explicit discussion of consciousness as especially mysterious and baffling goes back at least as far as the mid-nineteenth century. Thomas Huxley, the English biologist and prominent champion of evolutionary theory known as "Darwin's bulldog," famously wrote, "How it is that anything so remarkable as a state of consciousness comes about as a result of irritating nervous tissue, is just as unaccountable as the appearance of the Djinn, when Aladdin rubbed his lamp."[4] Eighty or so years later, that sentiment was echoed by Ludwig Wittgenstein in his *Philosophical Investigations*: "This is supposed to be produced in the brain! ... The feeling of an unbridgeable gulf between consciousness and brain process.... The idea of a difference in kind is accompanied by a slight giddiness—which occurs when we are performing a piece of logical sleight of hand."[5] Wittgenstein was skeptical about the reliability of this intuition, but he acknowledged its apparent power and its widespread acceptance in his day. A few decades later, in assessing the status of physicalism, Thomas Nagel expressed our sense of bewilderment about how matter might give rise to consciousness: "At the present time the status of physicalism is similar to that which the hypothesis that matter is energy would have if uttered by a pre-Socratic philosopher. We do not have the beginnings of a conception of how it might be true."[6]

If we come forward another fifty years to the present, does Nagel's assertion still hold? In this first quarter of the twenty-first century, does consciousness remain baffling in a deep and impenetrable way, or have the shrouds of mystery begun to lift in the face of scientific and philosophical developments? To what degree, and in what ways, if any, can we say that we currently understand consciousness and its basis? Has the problem of consciousness, like the question of life, moved from the status of a mystery to that of a research program? Consciousness is a complex phenomenon, and fully understanding it likely requires a diversity of theories and explanations, including theories at multiple levels, with interconnections between them. Thus, in assessing our current understanding of consciousness, we should not assume that success must involve a single eureka moment, one theory or insight that by itself resolves the puzzle. Perhaps there will be such a breakthrough, one that does for consciousness what the discovery of the double helix did for the problem of life, but perhaps not. The route to understanding may be more piecemeal and gradual, as many partial theories converge and collectively bring the nature of consciousness into focus.

In *Consciousness Reconsidered,* Owen Flanagan argued for a view of consciousness that he termed *constructive naturalism* and for an associated research program that serves as a paradigm example of the natural method in application.[7] Constructive naturalism is defined by three main claims: consciousness is real and important; consciousness is a feature of the natural physical world; and consciousness can be positively explained, meaning that with time, effort, and the right methods, we (humans) have the cognitive ability to explain consciousness by giving a satisfying, intelligible, positive explanation of *how consciousness exists as part of the natural physical world.* Formulating constructive naturalism in that way highlights how it contrasts with the various contemporary opponents that Flanagan's book aimed to challenge: eliminativists, dualists, and nonconstructive naturalists.

Flanagan's first target was the eliminativist view of consciousness. Conscious eliminativists resist the first element of constructive naturalism by denying the very existence or reality of consciousness.[8] They argue that there is no such thing, at least not if consciousness is understood in any way like the traditional notion of consciousness that generates the supposed mystery.[9] At first blush, the proposal seems extremely unpromising. It denies the existence of the one thing of which traditional philosophy held we were epistemically most certain—our own conscious minds, introspectively known from our individual internal perspectives. Despite the view's prima facie implausibility, eliminativists have various means by which to try to undermine that resistance and the intuitions on which it rests. On the one hand, eliminativists typically limit the scope of their denial. They concede that consciousness understood in some other less problematic senses may be real (e.g., consciousness as attention or metacognition), but they claim that those sorts of conscious phenomena are open to naturalistic explanation and do not generate any deep metaphysical puzzles or mysteries.[10] Thus, according to the eliminativist, we can still talk about consciousness, both in everyday life and in the lab. We just need to be clear what we mean by it, and we would not be giving up much that is of real value. We could retain the notions of consciousness that can do useful explanatory work while eliminating only the problematic notions of consciousness that supposedly result from mistaken philosophical theorizing. On the other hand, eliminativists also aim to undermine our standard intuitions about consciousness, by trying to show that they are the result of illusions or confusions. They often appeal to so-called intuition pumps, imagined cases designed to show us how our traditional ways of thinking about consciousness mislead us. Their aim is to alter our prior intuitions about consciousness and persuade us to discard our traditional notions of consciousness as incoherent or internally inconsistent.[11] According to the eliminativists, there is no mystery of consciousness. Our puzzlement results not from anything real in the world but only from our own confused notions of consciousness, which we should abandon. Eliminativism continues to be an actively supported view about consciousness, as seen in the recent advocacy of so-called illusionism.[12]

Flanagan's second and third targets were two forms of what he labeled "mysterianism,"[13] both of which accepted the reality and importance of consciousness but for separate reasons denied that we would ever be able to develop a positive naturalist physical explanation of consciousness. Mysterians of the first sort are mind-body dualists who argue that such an explanation is impossible because consciousness is not in fact physical.[14] Depending on the particular version of dualism involved, the reality of consciousness is said to require the existence of nonphysical substances (substance dualism) or nonphysical properties (property dualism) or require the view that everything real, every subatomic particle, has a mental aspect (panpsychism). Since the relevant substances or properties fall outside the realm of the strictly physical, we cannot give a positive physical explanation of consciousness.

Mysterians of the second sort concede that consciousness is physical—or at least may be physical—but argue that we humans lack the cognitive ability to construct and understand a positive physical explanation of consciousness. The second form of mysterianism counts as "nonconstructive physicalism" in Flanagan's terminology, because it asserts that we lack the ability to construct a positive intelligible explanation of consciousness and of how it is produced or constituted by its physical basis.[15] Any attempt to move from the physical to consciousness supposedly confronts what Joseph Levine has famously called "the explanatory gap."[16] We may be able to establish robust empirical correlations between physical or neural facts and various conscious facts, facts about the presence or absence of consciousness or about particular correlated features of conscious experience. We may even have good evidence that those correlations are nomic or law-like. However, according to Levine, we remain unable to give an explanation that makes intelligible why those correlations hold or any explanation of how the relevant physical facts might constitute or realize the correlated conscious features. The relevant gap is epistemic but not necessarily metaphysical or ontological, though some have tried to infer a metaphysical gap from the existence of the epistemic one.

Mysterians interpret the explanatory gap in a variety of ways, especially with regard to the possibility of its eventually being bridged. It might be asserted merely as a claim about the current state of our understanding, more strongly as a claim about the cognitive limits of any human understanding, or even more strongly as an *in principle* universal claim about the logical impossibility of any positive intelligible explanation of how physical facts might produce conscious experiences or constitute a conscious mind.

Claims of the last, very strong sort typically appeal to some radical differences between the items on the two sides of the psycho-physical gap to explain why it is impossible to bring them into an intelligible relation. In his seminal paper "What Is It Like to Be a Bat?," Thomas Nagel famously appealed to the difference between the "objective facts" of the sort that appear on the physical side of the gap and the "subjective facts" about experience that appear on the conscious mental side.[17] Subjective facts essentially involve an experiential point of view; they can be fully understood only from a specific experiential perspective. Only a bat (or other creature with similar echo-based sensory experience) can understand what it is like to be a bat. To understand the phenomenal experiences of a bat, one needs to be able to take the bat-type point of view oneself. In that sense, facts about experience are subjective; the ability to understand them is limited to a particular type of experiential point of view. By contrast, physical facts seem to be paradigmatically objective facts, facts that can be fully understood from many different points of view. Humans and extraterrestrial beings with very nonhuman systems might not be able to understand each other's conscious experiences in the what-it-is-like sense, but there seems to be no reason why they could not understand exactly the same physical facts and theories. Physical facts, unlike conscious facts, are objectively accessible from a wide range of experiential points of view. The supposed inability to logically bridge the subjective-objective divide thus provides one possible means to support a strong universal reading of the explanatory gap claim. Given the very different types of facts on the two sides of the gap, and the very different forms of understanding that are involved with each, it is impossible to understand their relation in an intelligible way; they cannot be logically combined in a way that lets us see intuitively how the one constitutes the other.

Other mysterian defenders of nonconstructive physicalism have appealed to more specific facts about the nature of human cognition to support their view that humans will never be able to close the explanatory gap, while leaving open the possibility that creatures with other sorts of cognitive abilities might be able to do so. For example, one of Flanagan's targets was Colin McGinn, who appealed to the nature and limits of human concepts and concept-forming abilities to argue that humans would never be able to understand the psycho-physical link.[18] According to McGinn, our physical concepts are inherently spatial and not of the right type to intelligibly explain consciousness. Moreover, McGinn also argued that the processes by which humans develop new concepts could never produce concepts of a sort that could make the psycho-physical gap intelligible. McGinn's claim was specifically a claim about the limits of human cognition. He accepted that consciousness has a fully sufficient physical basis, and he allowed that other beings with intellects and concept-forming abilities unlike our own might be able to intelligibly explain the nature of the psycho-physical link, but he concluded that we humans would never be able to.[19]

Thus, Flanagan's aim in arguing for constructive naturalism was to refute the eliminativists by showing that consciousness is real and significant, and to answer the mysterians by arguing that constructing a positive naturalistic physical account of consciousness is possible and that by following the correct methods we should expect to succeed in doing so, or at least that no sound arguments had been given to believe otherwise.

In *Consciousness Reconsidered,* Flanagan proposed a naturalistic method for studying consciousness that drew on three main approaches and sorts of evidence that should be explored concurrently in a mutually interdependent way.[20] The first was the investigation of the phenomenology of experience, especially as observed from the first-person point of view. In studying consciousness, it is essential that we begin with the intimate knowledge of experience that we acquire from our introspective awareness of its properties, structure, and organization. Introspection need not and should not be regarded as an infallible guide to the deep nature of conscious experience, nor should we assume that it provides us with fully correct information even about the apparent surface properties of consciousness. Nonetheless, our direct awareness of our own conscious experience provides our fundamental and reference-fixing grasp on what it means to be conscious. We know it most directly and essentially from our own first-person perspectives. Flanagan also argued for the value of attempts to understand the structure of experience more theoretically and to use more disciplined methods of self-observation, including those associated with the tradition of phenomenology.[21]

The second methodological approach and source of evidence was that of empirical psychology, especially research that aims to construct models of internal processing based on behavioral and performance data. As we will see, applying those methods to consciousness consists, in part, of attempting to gather accurate data about the differences, if any, between conscious and nonconscious mental processing. Such research may provide us with accurate models of the functional organization of consciousness and of the kind of processing that it involves. However, there is disagreement about the degree to which such functional models, even if correct, could count as fully explaining consciousness. Everyone agrees that the functional and information-processing aspects of consciousness are important, but not about whether those aspects capture the essence of consciousness. Some philosophers, such as Daniel Dennett, deny that there is anything more to consciousness than what can be captured in such functional theories. He attributes the belief that more is required to our having fallen victim to the illusion of what he calls "The Cartesian Theater."[22] For example, in response to those who claim an adequate explanation of consciousness requires an explanation of qualia, Dennett argues that there are no qualia to be explained. He is an eliminativist about the existence of qualia in any sense that corresponds to traditional philosophical use of the term *qualia*.[23] Others, however, argue that the essential aspect of consciousness is logically independent of its functional organization. David Chalmers argues that functional duplicate zombies are conceivable and indeed possible (i.e., it is at least logically possible for there to be functional zombie duplicates who have the very same functional organization as conscious human beings but who are themselves totally devoid of consciousness).[24] Those zombie duplicates process information just as conscious humans do, but there is nothing that *it is like to be them*. They have no phenomenal awareness of the sort we know from our first-person perspective. The possibility of such zombies is of course disputed, and thus there is no consensus about the explanatory power of functional models or their limits. Those who accept the possibility of functional zombies will deny that functional organization captures the essence of consciousness, but others will deny the possibility of functional duplicate zombies because they hold that functional organization does capture the essence of consciousness.

In addition to first-person phenomenological methods and behavioral and performance-based psychological theorizing, the third leg of Flanagan's overall research program was that of empirical neuroscience. Of particular importance was the need to investigate the neural substrates of consciousness, aiming to identify possible explanatory links between those substrates and the functional or phenomenological features identified by the two other legs of the natural method. If one aims to give a naturalistic and physicalist explanation of consciousness, one must ultimately show how the higher-level features of consciousness might be realized in neural structures and processes.

Flanagan was optimistic about the prospects that such a combined three-part research program would eventually lead to a satisfactory naturalistic explanation of consciousness. Each component would not only contribute to the overall theoretical explanation but would do so in a complementary and mutually interactive way, with each in continual dialogue with the other. A quarter century later, we can ask: Has research on consciousness been pursued in the ways that Flanagan advocated? And, if so, how has such research developed? What methods have been used? What issues have arisen? What questions have been asked? What results have been achieved? And to what degree, if any, do those results help to dissolve the mystery of consciousness? Is there reason to believe that we will soon be able to explain how consciousness depends on the brain just as we can explain how life depends on DNA and the rest of our molecular biochemistry?

The answer may depend on how similar the two cases—the case of life and that of consciousness—are. In particular, are the sorts of explanations required in the two cases of the same basic type, or does consciousness require a fundamentally different sort of explanation? Emphasizing the similarity, many philosophers and neuroscientists believe it is just a matter of time—and not too much time—until the mystery of consciousness is solved and nonphysical views of consciousness are as dead as vitalism is today.[25] Others argue[26] to the contrary that the mystery of consciousness is fundamentally different from the problem of life and that the methods used for explaining life cannot be used for explaining consciousness, at least not for explaining the essential nature of phenomenal consciousness and how it might be brought about (or realized) by purely naturalistic physical structures and processes.

David Chalmers, for example, argues that consciousness and life are not analogous. In the case of life, the problem was to figure out how various functional capacities—growth, metabolism, reproduction—could be accomplished by physical-chemical processes or mechanisms.[27] The explanatory project has been to explain how one set of functions or processes might be realized or produced by the organized interactions of lower-level functions and processes. It is a matter of decomposing higher-level functional systems into systematic organizations of lower-level ones sufficient to realize the higher-level functions and processes—reproduction, growth, or metabolism. It thus operates within an explanatorily homogeneous framework, mapping functions to functions and causal structures to causal structures.

However, according to Chalmers, the essential and problematic nature of consciousness that needs to be explained is not a functional ability or capacity. It is something quite different in kind—the subjective quality of experience, the what-it-is-likeness of being conscious as understood from our first-person perspective.

Chalmers has famously drawn a distinction between the so-called hard problem of consciousness and various easy problems of consciousness.[28] The latter involve explaining various functional aspects of consciousness, such as the sorts of information processing associated with consciousness. They may be difficult problems in a straightforwardly scientific sense, requiring lots of research and investigation, but they do not present any deep metaphysical or epistemological puzzles. They do not seem like mysteries, and we can imagine ways in which they might eventually be solved. By contrast, Chalmers's hard problem is that of explaining how qualitative phenomenal consciousness could be the result of purely physical processes. The explanandum and the supposed explanans involve fundamentally different aspects of reality—the what-it-is-likeness of being conscious as understood from the first-person perspective and objective structural and functional facts of the brain as understood from the third-person perspective of neuroscience. Thus, it is difficult to see any way in which the neuroscientific processes could suffice to produce phenomenal consciousness. It is this "hard problem" that generates an air of mystery and leaves us feeling perplexed and unable to see any way forward.

How robust is Chalmers's distinction between the hard and easy problems? In particular, are the two sorts of problems as distinct and independent as he claims? Chalmers relies on the assumption that making progress on the easy problems, such as questions about the functional organization and operation of conscious mental processes, will not be of much help in solving the hard problem. But is that assumption correct? Even if one concedes that there is at present a major unbridged explanatory gap, that does not preclude the possibility that future functional models may provide us with crucial insights into how the essential nature of consciousness might be realized. Indeed, it seems likely that understanding what consciousness *does* and what difference it makes to processing might help us understand what consciousness *is*. In other words, we might come to understand its essential nature by asking how that nature might enable it to function in the ways it does. It would be surprising if the essential nature of consciousness were epiphenomenal with respect to its functional features. As Flanagan argued in *Consciousness Reconsidered,* it is implausible to suppose that a structure as complex and biologically "costly" as consciousness would have evolved unless it had survival value, but it could not have any such value if it were epiphenomenal and made no difference to function. Even if its functional aspects do not suffice to capture its essential nature—which itself remains an open question—understanding its functional properties seems likely to help us determine what sort of essential nature could underlie such properties. In fact, given Flanagan's three-part research program, we should expect just such mutual interactions between our phenomenological analyses and our psychological processing models, and between each of them in turn with our neuroscience models.

As noted, the special divide that supposedly exists in the case of consciousness is sometimes described as that between the subjective and the objective.[29] Facts about conscious experience—about what-it-is-likeness—are said to be subjective in that they can be fully known or understood only empathetically by being able to imagine the relevant quality of experience; that is, by being able to imagine what it would be like to have that experience oneself. Doing so requires having a sufficiently similar subjective point of view, one that allows for the experience of the relevant quality. A congenitally blind person cannot understand the conscious experience of red in this empathetic sense because his or her subjective point of view does not include in its phenomenal repertoire the capacity to have such experiences. Thus, it is argued that no theory of objective facts of the sort that one might get from neuroscience or cognitive psychology can provide a full or sufficient understanding of subjective consciousness.

We can see this as yet another instance of the basic schema we associated with our three other mysteries: how can you get things that are F from things (or parts) that are not F? In this case, the specific question is: how can you get something that is *subjective* from things (or parts) that are *not subjective* (that are objective)? Indeed, that particular how-question may not have a positive answer. Thus, the even more basic questions are: *Is it possible* to get something *subjective* from things (or parts) that are *objective*? In other words, can any conditions specified in objective terms suffice to logically or nomically entail the existence of subjective facts?

In assessing that question, it is worth noting that the existence of subjective facts would seem to require the existence of a subject. Subjective facts are facts about how things appear from a particular conscious point of view, the view of some particular subject. An experience exists only insofar as it is the experience of some subject who is having the experience. For example, a pain can exist only insofar as some subject feels that pain. Experiences are perhaps best thought of as modes of subjects, as particular ways in which it is like something to be a conscious subject. Experiential states are states of subjects, states that subjects undergo rather than ones they merely observe. If experiences and subjective facts can exist only if there is a conscious subject to have them, then it follows that the conditions required to produce subjective facts would have to suffice to produce a conscious subject. Therefore, the moral to be drawn is that if you want to construct subjective facts, you have to construct the required subjects to instantiate those facts; to produce conscious experiences, you must construct the subjects who have those experiences. Failure to pay attention to that dependence may help explain why consciousness seems so mysterious.[30]

With regard to Flanagan's prediction, has a science of consciousness developed along the lines of the integrated research program he advocated? Certainly, consciousness is recognized today as a real and important part of mind and has emerged as a serious object of scientific investigation. Just a few years after the publication of *Consciousness Reconsidered* in 1992, two ongoing annual series of major scientific conferences were initiated. In 1994, the first meeting of The Science of Consciousness series was held at the University of Arizona, with large interdisciplinary conferences being held every year since then, either at the University of Arizona in even years or at a wide variety of international locations in odd years. Three years later, in 1997, the first meeting of the Association for the Scientific Study of Consciousness (ASSC) was held in Southern California, and the ASSC has held a major annual meeting ever since, attracting hundreds of psychologists, neuroscientists, and philosophers, with its most recent meetings being in Beijing in 2017, Krakow in 2018, and London, Ontario, in 2019. The number and quality of the presentations at those meetings reflect the explosive growth in research on the topic. An ever-increasing community of investigators have applied rigorous scientific methods to consciousness, articulating its structure and exploring its possible neural substrates.

Has this burst of research engaged in the three-part strategy of Flanagan's natural method? Perhaps the greatest single development that has driven consciousness research occurred at almost the same time that *Consciousness*

Reconsidered was published, for it was in 1992 that the first functional magnetic resonance imaging (fMRI) studies of brain activity using the blood oxygenation level dependent (BOLD) signal were published.[31] fMRI provided a noninvasive method by which changes in neural activity across different areas of the brain could be observed and measured in real time, allowing the collection of data and correlating selective increases or decreases in activation with various specific types of conscious and nonconscious processing. The primary technique uses the BOLD signal, which relies on the fact that the iron-containing hemoglobin in blood interacts differently with the intense magnetic field of the MRI scanner depending on whether the hemoglobin is oxygenated or not. Thus, the BOLD signal serves as a reliable indirect indicator of patterns of neural activity. It does so by identifying changes in blood flow and oxygen consumption in various brain regions as the relevant neurons increase or decrease their firing rates and their metabolic activity. The technique has a high spatial resolution and a moderate temporal resolution. Neural studies using fMRI are ubiquitous today, with more than one thousand such studies being published on average each month. A number of other noninvasive methods have been developed, such as magnetoencephalography (MEG) and enhanced electroencephalography (EEG), giving researchers the opportunity to observe the relevant patterns of brain activity using multiple techniques with different spatial and temporal resolutions, allowing them to corroboratively cross-check one method against others and build up a comprehensive model of the spatial and temporal dynamics of the neural processes underlying conscious mental activity.

These powerful imaging techniques have enabled research using a "method of difference" to perform comparison and contrast studies. The basic method is to compare cases of conscious processing with cases of nonconscious processing, otherwise keeping cases as similar as possible. The aim is to isolate the distinctive signatures of consciousness. For example, how do cases of conscious visual perception differ from unconscious visual perceptions of the same content? Such contrast studies can discover how conscious and nonconscious processes differ both functionally and neurally.

Many specific techniques are used to conduct such contrast studies. Perhaps the simplest is to use stimuli that are just above or below the threshold for conscious detection. Once the relevant threshold has been established for a given subject—typically by their reporting the conscious experience by verbal report or by pressing a button—then stimuli can be presented either above or below it, and neural and functional differences between the respective cases can be observed or inferred. Among other common experimental techniques for doing contrast studies are those that employ backward masking[32] or binocular rivalry with attentional modulation.[33] In a backward masking paradigm, the subject is first exposed to a stimulus well above threshold for 500 milliseconds, followed by a brief exposure to the target stimulus for only about 50 milliseconds, which in turn is followed immediately by another long-exposure stimulus for another 500 milliseconds. The latter stimulus serves as the so-called backward mask preventing the brief intervening target stimulus from being consciously perceived. Nonetheless, other indirect behavioral measures can show that the masked stimulus was unconsciously perceived and that information about it was processed often at higher levels. For example, if the masked stimulus is a word, it can produce unconscious behavioral effects related to its meaning, not merely to its form or appearance, despite the fact that the subject is not consciously aware of having seen the word at all. The crucial role of the backward mask is worth emphasizing. A 50-millisecond stimulus is sufficient to produce a conscious perception if the brief stimulus is followed by a blank screen rather than by a mask. It turns out that the visual processing of the mask is essential for preventing the visual processing of the target stimulus, which provides important evidence for attempting to understand the underlying mechanisms.

We want to understand what is present in the cases in which the stimulus is consciously perceived but is absent in the masked cases. Imaging research shows that, in the conscious cases, neural activity not only moves forward through the visual pathway but also feeds back from later higher processing areas toward earlier lower-level ones. The general visual pathway is from the primary visual cortex (first area V1 and then areas V2–V4, all in the occipital cortex) and continuing on to the middle and late visual processing areas in the inferior temporal (IT) cortex. In the case of conscious perception, activation also moves forward from the IT to areas in the prefrontal cortex (PFC) associated with attention and executive function.[34] Importantly, in the case of conscious visual perception, there is recurrent processing feeding back activity across every stage of the total pathway. First, there is recurrent processing from V4 to V2 back onto V1, then back from the IT to primary visual areas, and finally recurrent feedback from the PFC to the IT. In conscious visual processing, there is not only an initial spike in activity during the processing of the stimulus as information about it moves forward but also a secondary spike in those same areas as recurrent processing feeds back down the visual pathway, maintaining and strengthening the representation of the stimulus. By contrast, in cases of nonconscious visual perception, most of the relevant recurrent processing is absent, as is any stimulus-driven activity in prefrontal areas. The masking stimulus interferes with and suppresses the recurrent activity that would otherwise make the initial stimulus consciously perceived.[35]

Binocular rivalry offers another powerful means for conducting contrast studies. Subjects are presented with two different images, one presented to their right eye and the other presented to their left eye. Under such conditions, subjects do not report seeing both images simultaneously, nor do they see a superimposed double image. Rather, their conscious visual experience alternates between the two images, first one alone and then the other. Experimenters can use image types that are known to be processed in specific cortical areas, such as images of faces, which are known to selectively activate the fusiform face areas (FFA),[36] or images of scenes, which selectively activate the parahippocampal place area (PPA).[37] The results show that both images are being fully processed by the visual system, even though only one at a time is being consciously experienced. Binocular rivalry thus provides an excellent means to isolate the distinctive neural signatures of conscious visual awareness, specifically separating it from the components and processes that are shared by both conscious and nonconscious visual perception. What neural changes occur as the subject's awareness shifts back and forth between the two images? What seems to be involved is increased activation in the relevant specialized processing area (e.g., increased activation in the FFA when the subject reports that her conscious awareness has shifted to the face image). Moreover, that increase in activation may be produced in part by the integration of the relevant local specialized activation within a larger pattern of global activation involving many different brain areas, a fact to which we will return when considering the Global Neuronal Workspace theory of consciousness.

As well as providing information about the respective neural correlates of conscious and nonconscious perception, contrast studies also help to identify the functional differences between conscious and nonconscious mental processes. For example, we find important functional differences between conscious and nonconscious visual perception. Information that is consciously perceived or remembered is available for a wider and more flexible range of uses and applications as well as being more integrated with a broader range of other information. Thus, in this sense, conscious mental states and their contents are more unified with other states, both functionally and informationally. Other data on the functional aspects of consciousness come from clinical observations of deficit cases, cases of patients in whom neurological damage has produced selective losses of psychological function. Such research investigates the functional deficits that result when selective brain damage leads to a loss of conscious experience in some domain (e.g., vision) while still allowing some degree of unconscious processing to continue, manifested in the patient's residual ability to partially process the relevant stimuli sufficiently to produce at least some correct responses to those stimuli. For example, patients suffering hemineglect may fail to consciously perceive stimuli presented in their left visual field, though indirect measures may still show evidence that those stimuli have been processed to a high degree. By considering the differences in how conscious visual information can affect behavior versus how nonconscious visual information can do so, researchers can help identify the distinctive functional characteristics of conscious processing.

However, the significance of such research has been challenged by philosopher Ned Block, who has argued that it does not necessarily show anything about the function of conscious experience in the problematic what-it-is-like sense.[38] Block draws a distinction between two types of consciousness, which he terms phenomenal consciousness (p-consciousness) and access consciousness (a-consciousness). Information is access conscious in Block's sense if it is available for verbal report and for the flexible, rational control of behavior. By contrast, he defines phenomenal consciousness in terms of the presence of subjective experience in the what-it-is-like sense. He argues that in many contrast cases, including those involving neurological deficits, what is absent is access consciousness. The impact that such an absence has on behavior and on the operation of various mental systems is evidence only about the function of access consciousness. Block argues that phenomenal consciousness may still be present in some cases in which access consciousness is absent. Block argues that the two forms of consciousness need not always coincide and that indeed they in fact sometimes come apart. He argues that phenomenal consciousness may still be present in some cases in which access consciousness is absent. He gives the example of becoming aware that one had been phenomenally hearing a clock ticking for some period of time even though one was not conscious in the access sense of that stimulus. For the converse dissociation, he appeals to the possibility of a zombie who is functionally equivalent to a normal human with regard to access consciousness but lacks any phenomenal consciousness. Thus, according to Block, many contrast studies, including many neural deficit studies, do not provide evidence about the function of phenomenal consciousness. He argues that the failure to attend to the distinction between the two forms of consciousness has led to mistaken claims about the supposed functions of phenomenal consciousness, functions that in fact apply only to access consciousness. Interestingly, Flanagan made a reply to this line of argument in *Consciousness Reconsidered* before Block had published his seminal 1995 paper. Flanagan cites Block as having raised the objection to him in conversation. Flanagan's rejoinder challenges Block's basic distinction between access consciousness and phenomenal consciousness. He argues that the two are not in fact as distinct as Block supposes and that access and phenomenality are more closely connected than his critique supposes. Block concedes that the two forms of consciousness typically co-occur and that there may be significant links between them, but he argues that any such links must be independently established rather than assumed.[39] Debate on the issue continues.

Methodological questions have also been raised about the use of contrast studies to try to identify the neural substrates of conscious states and conscious processing in the brain. Using a variety of imaging techniques, including not just fMRI but also PET (positron emission tomography) scans, MEG, and enhanced EEG, researchers have tried to identify the neural differences associated with conscious mental states and conscious processing as opposed to those associated with nonconscious mental states and processing. Following the seminal terminology of Francis Crick and Christof Koch, such research is said to be aimed at finding the "neural correlates of consciousness," or NCCs.[40] However, the interpretation of such NCC studies has recently been challenged on methodological grounds.[41] In particular, the fact that the observed brain differences in the contrast case merely correlate with the presence of consciousness has led some to question whether it is appropriate to identify those observed states as the actual substrates of consciousness as opposed to merely its accompaniments. The limits on the evidential value of correlations are well known; any elementary statistics class stresses the fact that the temporal correlation of factors is not sufficient to establish a causal link between them: correlation does not prove causation. The same is also true of correlation and realization. The fact that neural features systematically correlate with the presence and absence of consciousness does not justify us in concluding that those neural features provide the substrate or realization base of the relevant conscious states.

Melanie Boly has argued that there are at least two alternative hypotheses that might explain the observed correlations.[42] On the one hand, the correlated neural features might be essential precursors of consciousness; they might serve as causes of consciousness that precede it rather than as the substrate that realizes or constitutes it. Unlike causes that precede their effects, realization and constitution are relations that are supposed to hold simultaneously between conscious states and their neural substrates. Discovering that the observed NCCs were the typical (or even nomically necessary) causes of consciousness would be an important result, but it would not be the same as finding the constitutive neural substrates of consciousness. Similar problems also arise in the other temporal direction. The observed NCCs might be regular successors or effects of consciousness rather than its substrate. For example, if conscious mental processes give rise to the ability to verbally report on one's mental state, the observed NCCs might be those that serve as the substrate of the relevant input to the language areas rather than as the basis of the conscious mental state itself. If we use a reportability criterion to determine the presence of conscious states, then conscious states will necessarily correlate with the presence of the neural states that underlie such reportability, but it does not follow that those states are part of the substrate of consciousness itself. It should eventually be possible to resolve these issues and determine which NCCs are genuine substrates and which are either causes of or effects of conscious processes, but at present care must be taken in interpreting contrast studies to identify NCCs.

The first-person phenomenological aspect of Flanagan's natural method has also developed in the past two decades, but again there has been controversy over fundamental issues. Going back to the origins of scientific psychology in the late nineteenth century, there has been a persistent problem of intersubjective disagreement about the features of phenomenal experience as reported by different introspective observers. Given the inherent privacy of introspective observations, it has been difficult to resolve such controversies. More recently, the problem of conflicting first-person judgments has led to debate about the range of properties that are manifest in phenomenal experience as well as about which sorts of mental states have a proprietary phenomenology (i.e., a phenomenology that distinguishes them as states of the type that they are).[43] With respect to the first of these issues, recent debate has focused on the phenomenal "thinness" or "thickness" of experience. For example, are the properties manifest in visual experience restricted to mid-level physical properties, such as color, shape, size, and motion, as some have argued,[44] or are more abstract and conceptual properties also manifest in experience. If I look at my smartphone in its brown leather case, is it only the phone's color and shape that are phenomenally manifest in my visual experience, or is the content of my awareness of it as my phone also phenomenally manifest? Those who favor a thick view of phenomenal experience will argue that my awareness of it as *my phone* is part of what-it-is-like for me to be visually conscious of it,[45] but those who support a thin view of experience will assert that any such contents that go beyond the relevant midlevel physical properties are not phenomenally manifest in my visual experience or in my general experiential state. Those latter supporters of thin phenomenology will concede that I recognize it as my phone and that this awareness is part of my current mental state, but they deny that such properties are phenomenally present in my visual experience. My visual experience of my phone will not differ from that of another person looking at it from the same angle, even if that person was unfamiliar with smartphones. Supporters of thick phenomenology by contrast would contend that my conceptual recognition of it as a smartphone does make a difference to my phenomenal experience and is a crucial part of what it is like to be me when I look at my phone.

A related controversy concerns which sorts of mental states have a proprietary phenomenology (i.e., a phenomenal character that distinguishes the state as being of the mental type that it is). There has been particular debate about whether cognitive propositional states such as believing, judging, or doubting the truth of some proposition p have a phenomenal character that essentially distinguishes them as states of that type (i.e., as beliefs or as beliefs that p). On this issue as well, there has been no consensus. Some have argued that such states themselves are not phenomenally manifest and there is no distinctive phenomenology of judging or believing, for example when I judge that Paris is the capital of France. They concede that when I occurrently make such a judgment, it may be accompanied by some other states that do have a manifest phenomenology, perhaps visual images of the Eiffel Tower, but they deny that the judgment itself has any phenomenology.[46] Supporters of that limited view of phenomenology are sometimes termed restrictivists, because they restrict proprietary phenomenal properties to perceptual and sensory states, including pain and pleasure states but explicitly excluding cognitive states such as judging, thinking, or doubting. However, others deny any such restriction. So-called expansionists recognize a far broader range of mental states as having proprietary phenomenologies.[47] According to expansionists, many nonsensory mental state types have a distinctive phenomenal character that is manifest to us in introspection. I can just know introspectively based on such a phenomenal character that I am believing that *p* versus thinking, wondering, or doubting that *p*. Expansionists claim that many such types of cognitive attitudes have a distinctive phenomenal character that is manifestly present to me in experience and that explains my introspective ability to know what sort of state I am in.[48] There is ongoing debate and disagreement about what aspects of our conscious mental states are phenomenally manifest.

Thus, even if we agree to define phenomenal consciousness as the sort that is involved in being something that "it-is-like-to-be," there is no consensus yet about what is and is not included in such what-it-is-likeness. In particular, there remains no agreement about what sorts of features are phenomenally manifest and thus must be included and explained as such in our theories. First-person introspection is an important source of evidence on such matters, but obviously it alone is not decisive, and larger theoretical considerations must also be taken into account in deciding what the phenomenological facts are. Though our theories must ultimately answer to the facts, theoretical considerations in turn often play a legitimate role in helping to determine what the facts are.

Turning to the second and third components of Flanagan's program, many psychological and neural theories and models of consciousness have been proposed and developed in the past quarter century. Two of the currently most influential approaches are the Integrated Information Theory of consciousness (IIT)[49] and the Global Neuronal Workspace (GNWS) theory.[50] Both models emphasize the extent to which consciousness involves the integration of information from multiple sources into a highly unified representation. The first defines consciousness in a highly general and abstract way relative to information theory. Giulio Tononi, the originator of IIT, constructs a mathematical measure φ to calculate both the information in a system's parts and the information contained in the organization of the system over and above that in its parts.[51] φ is thus a measure of the degree to which a system's information is integrated. According to IIT, such informational integration is both necessary and sufficient for consciousness and capable of being realized in a variety of physical substrates. Moreover, because informational integration comes in many degrees, IIT also allows for consciousness to come in many degrees, including very low degrees of consciousness that might be associated with a standard thermostat. If a system *S* has no subdivisions with φ values higher than *S*, then *S* will count as conscious according to IIT even if *S*'s φ value is not very high. It is just that *S*'s consciousness will also be of a very low degree. Thus, IIT appears to embrace a version of panpsychism, or at least very expansive and liberal psychism, that counts as conscious many systems that we do not ordinarily regard as conscious at all.

Global Neuronal Workspace theory perhaps comes closest to realizing the multipronged approach advocated by Flanagan, combining first-person data, functional models, and detailed neuroscience results about the correlates of conscious states. Initially proposed by Bernard Baars, Global Workspace (GWS) theory explains consciousness in terms of competition among processors for entry into a limited-capacity functional system that makes information broadly available through the brain for access and use by many different processors.[52] Being available in that way within the global workspace makes information conscious, at least in the access sense. It is available for report and the flexible control of behavior. The GWS model has been further developed and elaborated as the Global Neuronal Workspace (GNWS) theory, with proposed connections to particular neural and functional brain systems.[53] Of special importance is the claim that consciousness in both the access and phenomenal senses occurs when and only when the relevant content enters the larger global network involving both primary sensory areas and many other areas, including the frontal and parietal areas associated with attention.

According to Dehaene's GNWS theory, the same regions of the brain are involved in representing the contents of both conscious and nonconscious perceptions. For example, the contents of conscious and nonconscious visual perceptions are both represented by neural activations in the visual cortex. In both cases, the visual input is processed through the full visual pathway from early, to middle, to late stages of visual processing, moving from image analysis to the extraction of information about external geometric features and then to more abstract conceptual recognition of objects and individuals out in the visual scene. High-level conceptual categorization of objects can occur in nonconscious visual perception as fully as in conscious cases. In both cases, the path of processing activity moves from the primary visual cortex in occipital areas to a wide range of modular processing areas, primarily in the IT.[54]

In the GNWS model, the difference between conscious and nonconscious visual perceptions is a matter of how those processing areas in the visual cortex are interacting with other cortical areas. In the case of conscious perception, there are reciprocal activations between the visual cortex and areas of the prefrontal and parietal cortices that are associated with attention and executive control. Such looping recurrent activations between the visual and prefrontal cortices are not present in cases of unconscious perception. Moreover, in the conscious case, those interactions with prefrontal and parietal areas make the visual information widely available to many other diverse processing areas, to which they are also connected by long-range white fiber tracts. It is by that means that the information enters the global neuronal workspace and thus becomes conscious. The frontal and parietal activations also generate recurrent activation back onto the areas in the visual cortex that sent the signal forward. This serves to amplify and sustain the patterns of activation in the visual cortex associated with the initial representation of the visual stimulus.

Thus, in the case of conscious visual perception, there are two important and interconnected features that result from the feed-forward activation from visual to frontal and parietal areas. The representations in the visual cortex that generated the activations that were sent to frontal areas are strengthened by recurrent processing back from those areas, and those visual contents are made available to a wide network of other processing modules elsewhere in the brain through active long-range connections mediated through frontal and parietal areas. Dehaene sometimes refers to the transition from nonconscious to conscious as "ignition."[55] When the information moves from being nonconscious to conscious, recurrent feedback causes a significant increase in activity in the original processing areas where the representations of the stimulus still reside, as well as a general increase in global activation, especially in the relevant frontal and parietal areas. In a metaphoric sense, the relevant local and global circuits "light up," or "ignite," when the threshold from unconscious to conscious visual perception is crossed.

Dehaene's theory presents a promising and plausible model of consciousness that results from a research program very much in keeping with the three-component natural method proposed by Flanagan. First-person data play a key role. Dehaene primarily uses a subjective first-person report criterion to distinguish conscious cases from nonconscious ones. If a subject reports not having experienced the stimulus, the case is judged to involve only nonconscious perception, even if indirect behavioral measures indicate that significant visual information was extracted from the stimulus by the visual processing systems. Dehaene's theory also offers a detailed third-person psychological model GNWS that explains the functional changes that occur in global information processing when the relevant information becomes conscious by entering the global neuronal workspace. Also, imaging data provide evidence about which brain areas and which connections are involved in constructing the GNWS and what changes occur in the transition from nonconscious to conscious perception.

However, the GNWS model is not without critics. Appealing to Block's distinction between access consciousness and phenomenal consciousness, one can argue that the GNWS model provides at most an explanation of access consciousness, and leaves the deep mystery of phenomenal consciousness unsolved.[56] When information enters the global workspace, it becomes available to many processors and thus accessible for the direction of a diverse range of behaviors, including verbal reporting, exactly what is needed for access consciousness. But some have argued that entry into that global network is not required for phenomenal experience.[57] Phenomenal visual experience might occur purely as the result of activation in the visual cortex, perhaps within recurrent processing between late and early visual areas, but without any involvement of frontal or parietal areas. Such experiences realized wholly within the visual cortex would not be reportable by the subject, because of the lack of connection to language areas of the brain, but that need not entail that they do not occur. If one does not rely on a strict first-person reportability criterion for consciousness and shifts to an alternative objective behavioral or neural criterion, there may be reasons to believe that conscious phenomenal experiences occur in cases in which the subject does not report them.

Defenders of the GNWS can argue in reply that the notion of a conscious experience that the subject cannot report is incoherent since it seems to imply that the subject could have an experience of which the subject is unaware. How could a phenomenal experience occur without being present to some subject? Though the suggestion may seem contradictory, there are cases in which we have reason to believe that conscious perceptual experience is present despite the absence of verbal reportability. Split brain patients, for example, may provide some such cases.[58] In such patients, the corpus callosum has been surgically severed for medical reasons, and the two cerebral hemispheres no longer share information. Using carefully constructed laboratory displays with split screens, it is possible to provide different visual information to each hemisphere. Only the information presented to the left hemisphere is verbally reportable by the subject, given the lateralization of linguistic function to that side of the cortex. However, each hemisphere retains control of the contralateral hand—the left hemisphere controls the right hand and vice versa. Subjects can use their nondominant left hand to retrieve objects from behind a screen that matches the image or word that flashed on the left side of the screen and is being processed in their nondominant right hemisphere, despite the subject's sincere denial of having seen the stimulus. Given that the right hemisphere has no access to language processing areas, the subject cannot report what he or she saw in the left visual field. Yet the actions that the subjects make in response to that nonreportable information are of the sort that normally lead us to judge the subject as having had a conscious experience of the stimulus. The correct interpretation of so-called *split brain patients* remains a matter of controversy,[59] but such cases provide some reason to believe that phenomenal experience might sometimes occur in the absence of reportability, and if that can happen, then there is at least a possibility that activity within the visual cortex alone might suffice in some cases for phenomenal consciousness and phenomenal visual experiences. Any empirical evidence to support that hypothesis would be a definite challenge to GNWS theory. At present, it remains an open question.

Aside from the possibility of empirically based challenges, GNWS theory might be faulted at a deeper level if proposed as an explanation of phenomenal consciousness. Though the global functional and neuronal networks of GNWS seem well suited for explaining access consciousness, it is less clear how they might suffice to explain the subjective experience of a conscious self. Aside from the general mystery of the hard problem, there is a particular question about whether or how the systems to which GNWS appeals could have the sort of structure needed for phenomenal consciousness, particularly questions about how they might constitute the existence of a conscious subject of the sort required for phenomenal consciousness. The notion of phenomenal consciousness has traditionally been taken to refer to the way in which things *appear* in consciousness. By definition, the realm of the phenomenal is the realm of appearances, but how can there be an appearance unless there is some conscious subject or self to whom that appearance appears. An episode of appearing would seem to require a structure of *subject* and *object*—the object that appears and is present in experience and the subject to whom that object appears. The existence of the phenomenal object as present in experience essentially depends on its appearing as such to the experiential subject. Phenomenal objects obey Berkeley's classic dictum "esse est percipi," "to be is to be perceived."[60] The idea of an appearance without any subject to whom it appears is no more coherent than the idea of a pain that exists without being anyone's pain or felt by any subject. The principle that appearances require a subject to which they appear is the phenomenal version of the general principle that subjectivity requires a subject. Thus, if GNWS theory aims to explain consciousness, it must be able to explain how the neural functional global networks to which it appeals come to constitute or realize the existence of an experiential subject.[61] One might argue that the conscious subject is just the organism, since it is the organism that can be said to see, feel, and remember. Indeed, one might well identify the organism with the functional self (i.e., as the functionally organized system that processes information in all the relevant ways and acts appropriately in accord with it), but more needs to be said to explain how the organism might come to be an experiential self or subject. GNWS must specifically explain how the global neuronal network might constitute or realize an experiential self to which phenomenal appearances might appear.[62] GNWS theory may well be able to offer such an account in the future, but it has not done so yet. Thus, despite its power and promise, it is at present incomplete as a theory of phenomenal consciousness.

One very recent and new area of exciting consciousness research offers another exemplar of Flanagan's natural method, namely current investigations of the phenomenal, functional, and neural effects of various psychedelic drugs. For a variety of reasons, some of them social or political, there was for a period little serious research on psychedelics and consciousness, but recently there has been a resurgence of interest.[63] Psychedelics offer a powerful means of investigating consciousness by probing how various psychedelic drugs alter conscious experience, functional systems, and neural activity. When investigating any complex multilevel system, it is useful to have some way to perturb the system and observe how it responds at different levels, with a special focus on the correlations and interactions between the changes at the different levels. Such correlations can be of great value in constructing and testing models of how those different levels and aspects of the systems depend on each other. Psychedelic drugs can be used as probes to produce and observe such multilevel changes and interactions. Psychedelic drugs vary in the sorts of changes they produce in conscious subjective experience as reported by subjects from the first-person perspective both during and after the psychedelic experience. Different psychedelics also produce a wide range of functional changes in the operation of diverse psychological systems controlling cognition and behavior. Finally, psychedelic drugs differ widely in their neurological and neurochemical effects. Thus, by comparing the changes produced at all three of those levels, one can obtain critical data about how various aspects of consciousness and conscious experience depend on particular neural systems and types of neural activity, as well as insight into how those neural features underlie the sorts of functional capacities distinctive of consciousness. The use of psychedelics as a means of serious research in consciousness science is still quite recent, but the early results are very promising, and it may provide a critical source of evidence in the coming years.

In sum, we can conclude that Flanagan's 1992 naturalistic view of consciousness as a real feature of the world and a serious subject for scientific investigation has been strongly confirmed by the developments of the past quarter century. Consciousness studies has rapidly expanded as a field of scientific investigation, with a large number of active research programs worldwide employing a diverse range of empirical methods. Professional organizations, scientific journals, and major international conferences devoted to the scientific study of consciousness are well established. Many interesting theories and models have been proposed to explain consciousness and its possible neural basis. As in the examples given here, much of that research has been in line with the three-part natural method that Flanagan advocated, aiming to integrate first-person phenomenal data with functional models at a variety of levels and theories of the relevant neural structures and organization. Though the so-called hard problem of fully explaining how phenomenal consciousness might result from non-conscious physical or neural processes remains unsolved, the progress of the past twenty-five years has shown that Flanagan was right to be optimistic about the prospects for a constructive naturalistic explanation of consciousness and also right about the sorts of research that are likely to get us there. Perhaps by the end of the twenty-first century, the mystery of consciousness, like the earlier mysteries of life and the universe, will have been solved. The prospects look good, if less than certain.

Notes

1. Descartes, *Meditations on First Philosophy*.

2. James, *Principles of Psychology*.

3. Haldane, *Mechanism, Life and Personality*, 138.

4. Huxley, *Lessons on Elementary Physiology*, 193.

5. Wittgenstein, *Philosophical Investigations*, 124.

6. Nagel, What is it like to be a bat?, 436.

7. Flanagan, *Consciousness Reconsidered*.

8. See Churchland, Consciousness; Churchland, Reduction and the neurobiological basis of consciousness; Wilkes, Yishi, duh, um and consciousness.

9. Dennett, *Consciousness Explained*.

10. Rey, A question about consciousness.

11. Dennett, *Consciousness Explained*.

12. Frankish, *Illusionism as a Theory of Consciousness*.

13. Just as a historical aside, it is worth commenting on Flanagan's inventive and influential coinage of the term *mysterians* to characterize those who deny that any positive explanation is possible. Because the term so cleverly captures and frames the relevant set of views, it quickly became a viral meme in the field. When I once asked Flanagan about the source or inspiration for the name, he attributed it to the 1960s rock band Question Mark and the Mysterians, who enjoyed a brief period of fame with their breakout hit, "96 Tears." I myself recalled an earlier pop culture reference to a low-budget 1950s Japanese science fiction movie I had seen as a child entitled *The Mysterians*, about space aliens who invaded Earth before being defeated

by a newly united humankind. I was pleased when a bit of research revealed that the rock band had taken its name from the movie. Thus, the name was transmuted twice, from science fiction to rock and roll and then to philosophic meme, providing an interesting though arcane example of the natural method's openness to widely diverse sources of evidence and inspiration. Though the origin of the term might seem irrelevant to the intellectual merits or success of the claim, Flanagan's coinage—like Joseph Levine's "explanatory gap" or David Chalmers's "hard problem"—reminds us that the impact of theories sometimes depends on the psychological and cultural resonance of the viral terms in which they are expressed.

14. Chalmers, *The Conscious Mind.*

15. McGinn, Can we solve the mind-body problem?; McGinn, *The Mysterious Flame.*

16. Levine, Materialism and qualia; Levine, *Purple Haze.*

17. Nagel, What is it like to be a bat?

18. McGinn, Can we solve the mind-body problem?

19. McGinn, *The Mysterious Flame.*

20. Flanagan, *Consciousness Reconsidered.*

21. See Husserl, *Ideas*; Merleau-Ponty, *Phenomenology of Perception.*

22. Dennett, *Consciousness Explained.*

23. Dennett, Quining qualia.

24. Chalmers, *The Conscious Mind.*

25. Churchland, Can neurobiology teach us anything about consciousness?; Searle, *The Mystery of Consciousness*; Tononi, Consciousness as integrated information; Prinz, *The Conscious Brain*; Dehaene, *Consciousness in the Brain.*

26. Chalmers, *The Conscious Mind*; Strawson, *Consciousness and Its Place in Nature*; Gertler, In defense of mind-body dualism; Goff, *Consciousness and Fundamental Reality.*

27. Chalmers, *The Conscious Mind.*

28. Chalmers, Facing up to the problem of consciousness.

29. Nagel, What is it like to be a bat?

30. Van Gulick, Maps, gaps and traps.

31. Kwong et al., Dynamic magnetic resonance imaging of human brain activity during primary sensory stimulation.

32. Marcel, Conscious and unconscious perception.

33. Logothetis, Leopold, and Sheinberg, What is rivalling during binocular rivalry?

34. Dehaene, *Consciousness in the Brain.*

35. Ibid.

36. Kanwisher, McDermott, and Chun, The fusiform face area.

37. Epstein and Kanwisher, A cortical representation of the local visual environment.

38. Ibid.

39. Block, On a confusion about the function of consciousness; Block, Consciousness, accessibility and the mesh between psychology and neuroscience.

40. Crick and Koch, Toward a neurobiological theory of consciousness.

41. Boly et al., Are the neural correlates of consciousness in the front or in the back of the cerebral cortex?

42. Ibid.

43. Bayne and Montague, *Cognitive Phenomenology.*

44. Prinz, *The Conscious Brain.*

45. See Siewert, *The Significance of Consciousness.*

46. Prinz, *The Conscious Brain.*

47. Bayne and Montague, *Cognitive Phenomenology.*

48. Kriegel, *The Varieties of Consciousness.*

49. Tononi, Consciousness as integrated information.

50. Baars, *A Cognitive Theory of Consciousness*; Dehaene and Naccache, Towards a cognitive neuroscience of consciousness; Dehaene, *Consciousness in the Brain.*

51. Tononi, Consciousness as integrated information.

52. Baars, *A Cognitive Theory of Consciousness.*

53. Dehaene and Naccache, Towards a cognitive neuroscience of consciousness; Dehaene, *Consciousness in the Brain.*

54. Ibid.

55. Ibid.

56. Block, Consciousness, accessibility and the mesh between psychology and neuroscience.

57. Ibid.; Lamme, Toward a true neural stance on consciousness.

58. Gazzaniga, The split brain in man.

59. Schechter, *Self-Consciousness and "Split Brains."*

60. Berkeley, *A Treatise Concerning the Principles of Human Knowledge.*

61. Van Gulick, E pluribus unum.

62. Ibid.

63. Carhart-Harris, Leech, and Williams, Implications for psychedelic-assisted psychotherapy; Carhart-Harris et al., The entropic brain, 20; Icaza and Mashour, Altered states.

Bibliography

Baars, Bernard. *A Cognitive Theory of Consciousness.* Cambridge: Cambridge University Press, 1988.

Bayne, Tim, and Montague, Michelle. *Cognitive Phenomenology.* Oxford: Oxford University Press, 2012.

Berkeley, George. *A Treatise Concerning the Principles of Human Knowledge.* Dublin: Aaron Rhames, 1710.

Block, Ned. Consciousness, accessibility and the mesh between psychology and neuroscience. *Behavioral and Brain Sciences* 30, no. 5/6 (December 2007): 481–548.

Block, Ned. On a confusion about the function of consciousness. *Behavioral and Brain Sciences* 18, no. 2 (June 1995): 227–247.

Boly, Melanie, Massimini, Marcello, Tsuchiya, Naotsugu, Postle, Bradley, Koch, Christof, and Tononi, Giulio. Are the neural correlates of consciousness in the front or in the back of the cerebral cortex? Clinical and neuroimaging evidence. *Journal of Neuroscience* 37, no. 40 (October 2017): 9603–9613.

Carhart-Harris, Robin, Leech, Robert, Hellyer, Peter, Shanahan, Murray, Feilding, Amanda, Tagiazucchi, Enzo, Chivalo, Dante, and Nutt, David. The entropic brain: A theory of conscious states informed by neuroimaging research with psychedelic drugs. *Frontiers of Human Neuroscience* 8 (February 2014): 20. https://doi.org/10.3389/fnhum.2014.00020.

Carhart-Harris, Robin, Leech, Robert, and Williams, Tim. Implications for psychedelic-assisted psychotherapy: Functional magnetic resonance imaging study with psilocybin. *British Journal of Psychiatry* 200, no. 3 (March 2012): 238–244.

Chalmers, David. *The Conscious Mind.* New York: Oxford University Press, 1996.

Chalmers, David. Facing up to the problem of consciousness. *Journal of Consciousness Studies* 2, no. 3 (January 1995): 200–219.

Churchland, Patricia. Can neurobiology teach us anything about consciousness? Presidential address to the American Philosophical Association, Pacific Division.

Proceedings and Addresses of the American Philosophical Association 67, no. 4 (1994): 23–40.

Churchland, Patricia. Consciousness: The transmutation of a concept. *Pacific Philosophical Quarterly* 64, no. 1 (January 1983): 80–93.

Churchland, Patricia. Reduction and the neurobiological basis of consciousness. In *Consciousness in Contemporary Science*, edited by Anthony Marcel and Eduardo Bisiach, 273–304. Oxford: Oxford University Press, 1988.

Crick, Francis, and Koch, Christof. Toward a neurobiological theory of consciousness. *Seminars in Neuroscience* 2 (1990): 263–275.

Dehaene, Stanislas. *Consciousness in the Brain*. New York: Penguin, 2014.

Dehaene, Stanislas, and Naccache, Lionel. Towards a cognitive neuroscience of consciousness: Basic evidence and a workspace framework. *Cognition* 79, no. 1/2 (September 2000): 1–37.

Dennett, Daniel. *Consciousness Explained*. Boston: Little, Brown, 1991.

Dennett, Daniel. Quining qualia. In *Mind and Cognition*, edited by William Lycan, 519–548. Oxford: Blackwell, 1990.

Descartes, René. *Meditations on First Philosophy*. Paris, 1641.

Epstein, Russell, and Kanwisher, Nancy. A cortical representation of the local visual environment. *Nature* 392, no. 6676 (April 1998): 598–601.

Flanagan, Owen. *Consciousness Reconsidered*. Cambridge, MA: MIT Press, 1992.

Frankish, Keith. *Illusionism as a Theory of Consciousness*. Exeter, UK: Imprint Academic, 2017.

Gazzaniga, Michael. The split brain in man. *Scientific American* 217, no. 2 (1967): 24–29.

Gertler, Brie. In defense of mind-body dualism. In *Reason and Responsibility*, 13th ed., edited by Joel Feinberg and Russ Shafer-Landau, 359–372. Belmont, CA: Wadsworth, 2007.

Goff, Philip. *Consciousness and Fundamental Reality*. Oxford: Oxford University Press, 2017.

Haldane, John S. *Mechanism, Life and Personality: An Examination of the Mechanistic Theory of Life and Mind*. London: John Murray, 1913.

Husserl, Edmund. *Ideas: General Introduction to Pure Phenomenology*. Translated by William Boyce Gibson. New York: Macmillan, 1931.

Huxley, Thomas. *Lessons on Elementary Physiology 8*. London: Macmillan, 1866.

Icaza, Eduardo, and Mashour, George. Altered states: Anesthetics and psychedelics. *Anesthesiology* 119, no. 6 (December 2013): 1255–1260.

James, William. *Principles of Psychology*. New York: Henry Holt, 1890.

Kanwisher, Nancy, McDermott, Josh, and Chun, Marvin. The fusiform face area: A module in human extrastriate cortex specialized for face perception. *Journal of Neuroscience* 17, no. 11 (June 1997): 4302–4311.

Kriegel, Uriah. *The Varieties of Consciousness*. New York: Oxford University Press, 2016.

Kwong, Kenneth K., Belliveau, John W., Chesler, David A., Goldberg, Inna E., Weisskoff, Robert M., Poncelet, Brigitte P., Kennedy, David N., Hoppel, Bernice E., Cohen, Mark S., Turner, Robert, Cheng, Hong-Ming, Brady, Thomas J., and Rosen, Bruce R. Dynamic magnetic resonance imaging of human brain activity during primary sensory stimulation. *Proceedings of the National Academy of Sciences* 89, no. 12 (June 1992): 5675–5679.

Lamme, Victor. Toward a true neural stance on consciousness. *Trends in Cognitive Science* 10, no. 11 (November 2006): 494–501.

Levine, Joseph. Materialism and qualia: The explanatory gap. *Pacific Philosophical Quarterly* 64, no. 4 (October 1983): 354–361.

Levine, Joseph. *Purple Haze: The Puzzle of Conscious Experience*. Cambridge, MA: MIT Press, 2001.

Logothetis, Nikos, Leopold, David, and Sheinberg, David. What is rivalling during binocular rivalry? *Nature* 380, no. 6575 (May 1996): 621–624.

Marcel, Anthony. Conscious and unconscious perception: Experiments on visual masking and word recognition. *Cognitive Psychology* 15, no. 2 (April 1983): 197–237.

McGinn, Colin. Can we solve the mind-body problem? *Mind* 98, no. 391 (1989): 349–366.

McGinn, Colin. *The Mysterious Flame: Conscious Minds in a Material World*. New York: Basic Books, 2000.

Merleau-Ponty, Maurice. *Phenomenology of Perception*. Translated by Colin Smith. London: Routledge and Kegan Paul, 1962.

Nagel, Thomas. What is it like to be a bat? *Philosophical Review* 83, no. 4 (October 1974): 435–450.

Prinz, Jesse. *The Conscious Brain: How Attention Engenders Consciousness*. New York: Oxford University Press, 2012.

Rey, Georges. A question about consciousness. In *Perspectives on Mind,* edited by Herbert R. Otto and James A. Tuedio, 5–24. Dordrecht: D. Reidel, 1988.

Schechter, Elizabeth. *Self-Consciousness and "Split Brains."* Oxford: Oxford University Press, 2018.

Searle, John. *The Mystery of Consciousness*. New York: New York Review of Books, 1997.

Siewert, Charles. *The Significance of Consciousness*. Princeton, NJ: Princeton University Press, 1998.

Strawson, Galen. *Consciousness and Its Place in Nature: Does Physicalism Entail Panpsychism?* Thorveton, UK: Imprint Academic, 2006.

Tononi, Giulio. Consciousness as integrated information: A provisional manifesto. *Biological Bulletin* 215, no. 3 (December 2008): 216–242.

Van Gulick, Robert. E pluribus unum: Rethinking the unity of consciousness. In *Sensory Integration and the Unity of Consciousness*, edited by Christopher Hill and David Bennett, 375–392. Cambridge, MA: MIT Press, 2015.

Van Gulick, Robert. Maps, gaps and traps. In *Consciousness: New Philosophical Perspectives*, edited by Quentin Smith and Alexander Jokic, 323–352. Oxford: Oxford University Press, 2003.

Wilkes, Kathleen. Yishi, duh, um and consciousness. In *Consciousness in Contemporary Science*, edited by Anthony Marcel and Eduardo Bisiach, 16–41. Oxford: Oxford University Press, 1988.

Wittgenstein, Ludwig. *Philosophical Investigations*. Oxford: Basil Blackwell, 1953.

2 Domesticating Scrupulosity

Robert N. McCauley and George Graham

Introduction

William James wrote that "The facts of the world in their sensible diversity are always before us, but the philosophic need craves that they should be conceived in such a way as to satisfy the sentiment of rationality. The philosophical quest ... is the quest [for that] conception."[1]

Owen Flanagan has been on a creative philosophical quest throughout his long and distinguished career. He has written about a wide range of problems in philosophy of mind, ethics, moral psychology, science, the philosophy of mental illness, and religion. A distinctive and synoptic philosophical picture has emerged from his discussions of philosophical problems: a theory that combines realism about phenomenal consciousness with a multidisciplinary pluralism about scientific understanding, an account of the reason responsiveness of persons that acknowledges human susceptibility for mental disorder and distress, a tale of the human desire for religious transcendence alongside a skepticism about theism, and an ethics of human behavior with distinctive turns into the topics of the meaning of life and culturally situated moral virtues.

An explanatory pluralist methodology, in the form of his "natural method," plays a prominent role in Flanagan's analyses from topic to topic.[2] One of the benefits of explanatory pluralism, according to Flanagan, is that it can provide us with an analysis of what may be learned both from abnormal cases of a phenomenon, say, disorders of cognition or decision-making and from functional or normal cases of the same. We ourselves are independently attracted to a pluralistic conception of things and, like Flanagan, especially when it comes to matters of mind.[3] As Flanagan puts things, the method "consists of seeking consistency and equilibrium among different modes of analysis applied to the study of some mental phenomenon."[4]

This chapter embraces a pluralistic methodology as we examine a pathological form of religiosity known as scrupulosity and what we call its "domestication." The chapter gives us a welcome opportunity to help honor our dear and deeply admired friend and to try to sculpt a philosophic contribution that expresses some of the philosophic interests and sympathies that we share with him.

Religious Obsessions and Compulsions

Scrupulosity Abounding

The relentlessness of the moral and spiritual tacking by the seventeenth-century Puritan writer John Bunyan in his memoir, *Grace Abounding to the Chief of Sinners,*[5] is nothing short of dizzying. Bunyan wrote this autobiographical work during his twelve-year stay in the Bedfordshire jail as, arguably, a prisoner of conscience. (This is when Bunyan also began work on his best-known book, *The Pilgrim's Progress from This World, to That Which is to Come.*) He had joined a small congregation of Puritan nonconformists and eventually became a preacher. Technically, his preaching was in violation of laws aimed at ensuring the preeminence of the Church of England. The authorities' zeal for enforcing such laws, however, increased considerably with the restoration of the monarchy in 1660, when Bunyan was prosecuted. Bunyan was held because of his refusal to discontinue his preaching. For so steadfast and devoted a prisoner of religiously informed conscience, whose imprisonment left his wife and four children without any means of support, the spiritual life Bunyan portrays in *Grace Abounding* seems unexpectedly wobbly.

That impression is, no doubt, partly a function of the conventions of conversion narratives. Bunyan begins *Grace Abounding* by discussing his first three decades, including his service in his late teens in the parliamentary army in the 1640s during the first of the English Civil Wars. Of a piece with those narrative conventions, Bunyan establishes in the opening pages the moral waywardness of his youth, stating that as a child he had "few equals ... for cursing, swearing, lying and blaspheming the holy name of God." He also covers the requisite episodes when, his ungodliness notwithstanding, it appeared, retrospectively, that God had saved him from a premature death. Bunyan does comment, though, that throughout his youth he was regularly terrified by "fearful dreams" and "dreadful visions" of devils attempting to enlist him and that he was "greatly afflicted and troubled with the thoughts of the day of judgement ... fearing that it would be my lot to be found at last amongst those devils and hellish fiends."[6]

Bunyan indicates that it was during his twenties that his engagement with religious ideas and practices became simultaneously more serious and more fraught. He discusses numerous episodes that illustrate his spiritual vacillation. He also describes his many efforts to forgo the delights he took in what he had come to regard as problematic entertainments such as dancing, bell ringing, and playing games.

His account of the latter, in particular, is characteristic of the tenor of the remainder of the book. In an episode concerning game playing, he alternates between the highs of godly endeavor and the lows of sinful conduct twice in a matter of hours. He describes his brooding on the local parson's Sunday morning sermon about failures to honor the Sabbath day, but having shaken off the sermon's effect over lunch, that afternoon he returns to playing games. But, as noted, it does not end there. Bunyan then reports that in the midst of his afternoon sport he hears a heavenly voice questioning the fate of his soul. Instead of what is, presumably, an astonishing experience proving transformative and redirecting his life forevermore, he decides in fairly short order to charge ahead with his no longer so innocent merriment, reasoning that, "I had as good be damned for many sins, as to be damned for few."[7] His religious ardor had failed him by lunchtime, and not even hearing an admonishing heavenly voice could revive it in the afternoon.

Such spiritual oscillations are routine stuff in Christian conversion narratives. Driven by reliance on the law, belief in good works, pride, misunderstandings about Christian salvation, failures of faith, or doubt, humans fail to accept the grace of God. With sudden conversion, descending to some lowest of low points before finally accepting God's grace is fairly standard fare. Bunyan's conversion, however, is more gradual.

What is striking about *Grace Abounding* is Bunyan's recounting of what seem to be ceaseless rounds of expressions of surety in his salvation undone by some new temptation from Satan, inspired dedication subverted by some new (often biblically informed) recognition of error, or feelings of spiritual triumph followed by nearly instantaneous doubt. Even as Bunyan addresses and encourages his readers, he repeatedly acknowledges his struggles to escape his constantly recurring doubts, noting, for example, that temptations to betray Christ "sometimes ... run in my thoughts not so little as a hundred times together."[8]

Intense feelings of guilt or recurring doubts about one feature or another of the formula for salvation, especially arising so soon and so consistently after all too brief moments of religious confidence, are a characteristic symptom of what has come to be regarded as a pathological form of religiosity known as *scrupulosity*. It does not seem unfair to assert that, no less than grace, evidence of scrupulosity abounds in Bunyan's book as well.

Characterizing Scrupulosity

Scrupulosity, according to the fifth edition of the *Diagnostic and Statistical Manual of Mental Disorders* (*DSM-5*),[9] is a form of obsessive-compulsive disorder (OCD) combined with excessive conscientiousness.[10] In one memoir, an OCD patient drolly describes scrupulosity as "OCD plus Jesus."[11] We take up both scrupulosity and OCD in greater detail further on, but for now we simply wish to convey a sense of the scrupulosity that Bunyan's memoir seems to illustrate.

Instead of worries about contamination, physical security, or social well-being, which are salient concerns in other forms of OCD, it is disquiet about religious or moral matters that preoccupies the scrupulous. Usually they are tormented by both. A cognitive model of scrupulosity allows that "one's religious beliefs and values influence the misinterpretation of normally occurring unwanted thoughts as potentially sinful or foreboding of divine punishment."[12] Scrupulous individuals fret incessantly about committing religious or moral offenses or overlooking religious or moral obligations. As with other forms of OCD, scrupulous people often focus on matters that other (in this case) religious people regard as trivial (e.g., daydreaming during group prayer). The scrupulous neither distinguish between accidental and intentional violations of religious prescriptions nor mark their comparative importance. They see "sin where there is none" and live in unending fear of divine punishment for their lapses.[13]

That is largely because *they exhibit all of the other prominent patterns associated with OCD*. The scrupulous are entangled in a web of psychological dispositions that are mutually reinforcing.[14] Scrupulous individuals accord great moral and causal significance to their thoughts. They agonize about their inability to control their thoughts and, particularly, their problematic intrusive thoughts, which they do not want. All of that leads to constantly performing rituals and pursuing reassurance wherever it can be found, to relieve intense anxiety arising from their obsessive thoughts about sin and about failures to fulfill religious obligations. These measures are endlessly duplicated because their effects are quickly undone by the thoroughgoing "intolerance of uncertainty" by the scrupulous about what are, in this case, matters for which no certainty can be obtained.[15] Their characteristic, if misplaced, perfectionism forbids tolerating any doubts. Jonathan Abramowitz and Ryan Jacoby comment that "it is as if those with scrupulosity have *lost their faith in faith*."[16]

Pathological Religiosity or the Protestant Order of Salvation?

Exhibiting obsessions and compulsions with regard to religious matters has historically been termed as having "scruples," and exhibiting such scruples qualifies as *scrupulosity*, according to Joseph Ciarrocchi, when it satisfies the following conditions (enlisted from Greenberg):[17]

- The scruples' focus is comparatively narrow.
- The surrounding religious community regards the concerns as quite minor.
- The obsessions interfere with "normative religious practice."
- The subjects themselves find their scruples troubling.
- The scruples are accompanied by other symptoms of OCD.

Throughout *Grace Abounding*, Bunyan wrestles with what he regards as his habitual sinfulness, his persistent doubts, his constant worries about blasphemy, his incessant urges to engage in all these untoward thoughts and behaviors, and the recurring guilt associated with each that he feels.[18]

That dizzying frequency and the consistency with which such matters arise in *Grace Abounding* constitute ample evidence of the *intrusiveness* of these thoughts. Sometimes Bunyan ascribed these temptations and impulses to the influence of Satan,[19] but with all the moral, religious, and ritual expectations that many religions impose (there is nothing unique about Christianity here), they are fully sufficient on their own, without references to demons, to raise innumerable occasions for apprehensions.[20] *Grace Abounding* is crammed with biblical citations, on which Bunyan continually ruminates, but just as often he reports that these passages have "darted" to mind (i.e., they have intruded unbidden).[21] Considered reflectively, they may provide assurance, but as unwanted intrusions they, in effect, indict him for some additional way he has fallen short.

Bunyan's memoir also includes multiple instances of one of that pattern's characteristic consequences, his unending need to *seek reassurance* about his salvation.[22] For example, Bunyan states that scriptural affirmations of God's redemption "would make me make a little stop, and, as it were, look over my shoulder behind me, to see if I could discern that the God of grace did follow me with a pardon in his hand, but I could no sooner do that, but all would be clouded and darkened again."[23] *Grace Abounding* contains dozens of passages in which Bunyan reports recurrent cycling through moments of religious conviction undone by fears, disquiet, and doubts, only to have some further scripture, experience, reflection, or ritual offer some reassurance. Bunyan remarks that "my peace would be in and out sometimes twenty times a day."[24] Spirituality of this sort is not just exhausting; it is frequently depressing.[25]

Most people who suffer from OCD and are reflective eventually come to *recognize* many of the patterns characteristic of this disorder *as patterns*. Bunyan readily perceived his penchant for entertaining the same bothersome thoughts over and over again, his distress both about those thoughts and about having those thoughts, the various measures (seeking reassurance, carrying out rituals) that he undertook to ameliorate those negative effects, and the failures of those measures to provide any long-term relief from that distress. It seems unlikely that Bunyan was unclear about the fact that consciously recognizing these patterns, in and of itself, does not help eliminate them.

Whether Bunyan or other seminal figures in the history of Protestantism (e.g., Luther) are well understood as exemplars of scrupulosity as a form of *pathological* religiosity is not an uncontroversial matter. Paul Cefalu cautions against applying contemporary standards anachronistically. He holds that figures from Reformation Protestantism exemplified a "rather common form of excessive ... *non-pathological*, religiosity," which constitutes a crucial stage in what Cefalu calls "the Protestant order of salvation."[26]

This is born, in no small part, of the special prominence that the leaders of the Reformation attach to scripture. Luther championed the authority of the Bible over the pronouncements of the church. At various points in the Christian Bible, though, Jesus not only warns against sinful actions but also warns against sinful *thoughts*, declaring, for example, that a male who even looks at a female lustfully has already committed adultery with her "in his heart" (Matthew 5:28). He suggests that thoughts about bad actions can be as bad as the bad actions themselves. Thinking about doing something sinful and actually doing it carry comparable moral gravity. This is the moral variant of what is known as "thought-action fusion" (TAF). Roz Shafran and her colleagues define *moral TAF* as "the belief that having an unacceptable thought is the moral equivalent of carrying out the unacceptable or disturbing action."[27] The other variant of TAF is *likelihood TAF*, which is the belief that thinking about some untoward event makes it more likely to occur. With both forms of TAF, people's own thoughts drive their sense of causal and moral responsibility.

Cefalu maintains that for these early modern figures this religious obsessiveness fits "into a larger life-narrative, [as] a necessary means to a spiritual end."[28] Only after experiencing the futility of trying to follow God's commandments in every respect and thereby coming to appreciate their inevitable and thoroughgoing failure to do so do these individuals surrender themselves to salvation by grace alone, through faith alone, in Christ alone, on the basis of scripture alone.[29] Cefalu comments that for these Protestants "only an abdication of creaturely willfulness to impel one's salvation can prepare the individual to receive transformative bestowals of divine grace.... This ... stage is, counterintuitively, achievable through the very means of non-pathological obsessive religiosity."[30] It is only through realizing the ineffectiveness of their scrupulous attempts to follow every commandment, meet every religious and moral demand, and live a perfect life[31] that they come to "relinquish subjective agency.... Scrupulosity is the positive condition of its own negation and eventual supercession."[32] Their scrupulosity creates the opening, via their recognition of the hopelessness of their own efforts, for them to gain salvation by submitting to and receiving the grace of God. On the one hand, Cefalu emphasizes that this nonpathological Protestant order of salvation is the key to their ultimate spiritual success, though he acknowledges, on the other hand, that *nothing follows from that about their liberation from scrupulosity*. That concession is crucial. We will suggest, by contrast, that with regard to scrupulous individuals such as Bunyan (and probably Luther), this is better understood not as nonpathological religiosity but as what we call the *religious* "domestication" of (pathological) scrupulosity.

Scrupulosity across Religious Cultures

On the assumption that rates of OCD in the sixteenth- and seventeenth-century German and English populations did not vastly exceed those reported today, Cefalu's claim that the religious obsessiveness characteristic of scrupulosity was a "necessary means to a spiritual end" is hyperbolic. Not even most Protestants, then or now, suffer from scrupulosity. This would imply that alternative routes to that spiritual end were and are available to Protestants.

Behind Cefalu's overstatement, though, is a general point that is worthwhile. Different cultural circumstances may lead to different standards for what counts as pathological religiosity.[33] Norms about such matters, just like norms pertaining to religious people hearing voices—as Luhrmann's study demonstrates, may vary across cultures.[34] Something like the patterns in the mental lives and conduct of people with OCD that are idiosyncratic and profoundly troubling can prove normative and affirming within accommodating frameworks that religions provide.[35] Inquirers should be wary of false positives. Routine ritual performance can appear easily enough to an outsider to be scrupulous preoccupation. In short, culture matters.

In the case of early Protestantism, Cefalu proposes that it is the emerging understanding during the Reformation of scripture and Christian life that makes for religious arrangements that yield patterns associated with OCD. He comments: "What early modern non-pathological obsessiveness shows is that a culture's organizing worldview can create both the conditions under which obsessions and compulsions might emerge, as well as historically particular evaluative criteria that might depart from 21st-century measurements of OCD."[36]

This pertains directly to the third of Ciarrocchi's criteria of scrupulosity mentioned earlier, scrupulous obsessions' interference with "normative religious practice."[37] Religious representations and arrangements that result in the widespread cultivation of religious obsessions and compulsions in a population may well shift what counts as normative religious practice and thus the norms that delineate what counts as pathological religiosity as well. The sort of cultural domestication of scrupulosity that Cefalu proposes, however, *does not cure it*. His wavering assurance of his salvation notwithstanding, Bunyan remains a captive of his mental disorder. Instead, such domestication enables the scrupulous to proceed less conspicuously and even, perhaps, to go undetected in a larger population of participants whose mental lives and behaviors, as the result of the prevailing religious representations, mimic many of the features of the scrupulous.

We suspect that Cefalu's discussion of the impact of the Protestant order of salvation is only a beginning. In what follows, we will sketch a case for the broader application of Cefalu's speculation about the ability of some religious and cultural arrangements to encourage and therefore regularize religious obsessions and compulsions. The regularization of such thought and conduct, however, does not relieve the distress of the scrupulous.

First, research on *contemporary* Protestantism demonstrates the same patterns. Moral TAF plays a no less salient role in Protestant theology today than it did more than four centuries ago. Comparative research indicates that Protestant participants in the United States, Netherlands, and Belgium show greater levels of religiosity than do participants from other religious groups and that they are far more likely to subscribe to moral TAF.[38] Abramowitz and his colleagues provide evidence from a collection of self-report measures that highly religious Protestants not only exhibit a host of obsessional symptoms but experience them with significantly greater severity than do either Protestants of more moderate religiosity or agnostics and atheists.[39] Abramowitz and Jacoby also subscribe to the view that with at least some religious systems (including, presumably, numerous contemporary Protestant variants) scrupulosity can arise among those predisposed to OCD as the result of fervent religiosity within the pursuit of conventional religious forms.[40]

Nor among Christians are such religiously abetted obsessions confined to Protestantism. Although his ideological foe, Luther's Catholic contemporary Ignatius Loyola also constantly doubted his salvation. Like Luther, for example, he repeatedly confessed the same sinful incidents. Prior to the Reformation, early modern Catholicism was already well acquainted with scrupulosity. In the fifteenth century, theologians had invented the notion of an "erroneous conscience" to be applied to scrupulous individuals. That principle held that people manifesting scrupulosity *incorrectly* judged the moral and religious matters that burdened their religious lives, and the church allowed priests to exempt the scrupulous from "laws and practices required of the non-scrupulous."[41] Ciarrocchi argues that the development of these special measures furnishes plentiful historical evidence of the long-standing recognition of the troublesomeness of scrupulosity.

Scrupulosity is also observed among Muslims. Not only does Islam have warnings about sinful thoughts, the most prevalent versions of Islam also generally have more ritual requirements about purity and prayer than do the most prevalent versions of Christianity. In one recent set of studies, Muslims in Turkey exhibited higher levels of OCD symptoms than did the researchers' Christian participants in Canada for every domain tested, including moral TAF.[42] Studies of OCD across cultures also seem to indicate that OCD patients in majority-Muslim countries are more likely than OCD patients in other cultures to report specifically *religious* obsessions.[43]

Just like the evidence about scrupulosity in the Protestantism of the sixteenth and seventeenth centuries, in the end, this is unlikely to turn on anything *unique* about Islam. Abramowitz and Jacoby review how Christianity (both Protestant and Catholic), Islam, Judaism, and Hinduism each offer conspicuous resources for preoccupying the scrupulous.[44]

OCD and Cognitive Models Thereof

OCD

Although OCD encompasses a host of different behaviors, thoughts, and feelings, some commonalities are identifiable. People experience persistent, unwanted obsessions that provoke anxiety, which they resist by seeking relief through carrying out compulsive behaviors or thoughts (though either obsessions or compulsions can arise independently of the other). The obsessions are intrusive thoughts, images, or impulses that are ego dystonic (i.e., distressing). The compulsions are ritualized behaviors or thoughts that temporarily relieve such obsession-driven anxieties but that are detrimental in the long term.

Both the obsessions and the compulsions can take a variety of forms, which, at least collectively, appear to cut across cultures:[45]

> Obsessive-compulsive disorder symptoms have been classified into a number of common subtypes, including contamination obsessions with cleaning compulsions; accidental harm obsessions with checking compulsions; aggressive, sexual, and religious obsessions associated with compulsions to prevent negative outcomes; symmetry obsessions with ordering and arranging compulsions; and doubting or perfectionism obsessions with counting and repeating compulsions.[46]

Ciarrocchi holds that the range of possible materials populating those obsessions and compulsions is "infinite."[47]

OCD's potential for disruptiveness can be difficult to imagine. David Eilam and his colleagues supply a diary entry of a patient describing how turning on her television involves, *each time*, roughly a half hour of causally unrelated, repetitive, ritualized behaviors that she must go through.[48] To be burdened for years, often decades, by unrelenting obsessions and often complicated, time-consuming, and completely irrelevant compulsions that trouble every aspect of a patient's life can be profoundly depressing. OCD and major depression are comorbid, with estimates that as many as two-thirds of OCD patients experience at least one episode of major depression in their lives.[49] In extreme cases, patients have even resorted to suicide.

Earlier editions of the *DSM* classified OCD among anxiety disorders, but *DSM-5* accords OCD its own category.[50] Cognitive models have provided insight about how the contents of various OCD symptoms are linked to a range of psychological findings.[51] The problem is not that people with OCD experience unwanted intrusive thoughts. *That* is an ordinary phenomenon—estimated to arise in 90 percent of the population.[52] These cognitive models propose instead that the problems begin when people with OCD incorrectly assess the relative importance of their intrusive thoughts. They assume, first, that the risks to their or others' biological, psychological, or social welfare, to which these troubling thoughts pertain, are sound and, second, that they bear responsibility personally for any unhappy consequences, for themselves or for others, of failures to mitigate such threats. People with OCD also grievously misjudge the protection from such outcomes furnished by what are for everyone else *routine* precautions. This appears to arise from (perhaps in different circumstances) either a failure to recognize the successful completion of the precautionary measure or a failure to remember carrying it out. Perhaps not initially, but eventually, most people with OCD come to understand these patterns,[53] but, by itself, understanding them is not enough to bring about meaningful behavioral change.

Obsessions

Obsessions are not merely exaggerated worries about actual problems the person faces. They are unwanted, intrusive mental events, the experience of which individuals find discomfiting. Resources that have been counted do not require repeated recountings. Thoroughly clean things do not need to be thoroughly cleaned repeatedly. If five cleanings are not enough to address the nagging doubt about cleanliness, why would a sixth, a sixteenth, or a sixtieth cleaning suffice?[54] Obsessions may not even be about empirically possible risks. One memoir recounts the author becoming obsessed with the possibility that a scenario from a science fiction novel he had read might someday be realizable, its physical implausibility notwithstanding.[55]

The pitilessness of the anxieties associated with obsessions can, for a time, thoroughly obscure the bearing of reasons on those obsessions' pointlessness.[56] People with OCD often attach unmerited significance to their thoughts, including distressing thoughts that simply pop into mind. When thoughts are informed by good reasons, they *can* be important. The importance in question here, however, arises exclusively from people simply *having* thoughts that they find disturbing or believe violate some norm that they embrace. In OCD, nothing more than the fact that such a thought has *occurred* is the ground for its putative importance. People with OCD are not impervious to reasons. Arguably, they *overreact* to what are otherwise reasonable concerns. Because their reactions are so overwhelmingly driven by the mere occurrence of such thoughts, though, the scrupulous basically fail to be reason-responsive.[57]

For many OCD patients, their eventual recognition of the unreasonableness of it all only contributes to their disquiet. Sooner or later, they come to comprehend the senselessness of their fears and the absolute irrelevance from an instrumental standpoint of many of their compulsions.[58] Success at turning on a television set does not depend in any way on touching the curtain across the room, moving the lamp three times, and performing nine hand-washings in the previous thirty minutes.[59] Recognizing the irrationality of obsessive fears and the futility of compulsive rituals only adds to the sense of helplessness and anxiety that the person feels.[60]

Obsessions frequently are unwanted *thoughts*. OCD exploits the fact that thoughts are cheap and a good deal less controllable than we customarily presume. As one example goes, try not to think of an elephant. At any given moment, *most* people find it challenging to suppress particular thoughts or images on demand.[61] They find it especially difficult to subdue unwanted thoughts about perils. Attempting to suppress the thought seems as likely to endow it with greater vitality as to squelch it.[62]

Compulsions

Obsessions almost always engender compulsions, though perhaps as many as half of OCD patients have at least one compulsion without an associated obsession.[63] Still, careful clinical scrutiny usually uncovers a corresponding obsession.[64] People with OCD experience compulsions to act or think in ways that are stereotyped, which they inevitably repeat over and over, in response to some persistently recurring obsession. Repetition is such an obvious feature of OCD that some researchers have simply proposed measuring its severity by how much time the repetitious behaviors consume.

Repetitive behaviors or thoughts are compulsions when they result from an internal impetus to act or think in some way or other in response to an obsession, when they are undertaken in order to reduce the disquiet associated with that obsession or to prevent some worrisome outcome, and when, in less fraught contexts, the agent judges those actions as "excessive, exaggerated, and ... senseless."[65] The six most frequent compulsions are checking, washing, counting, asking or confessing, establishing and preserving symmetry and precision, and hoarding,[66] though hoarding is classified as an identifiable disorder itself in *DSM-5*. Compulsive acts constitute personal rituals that are rigid, stereotyped responses to obsessions. Experimental comparisons of OCD patients and controls carrying out various ritualized motor tasks, such as filling a pet's water bowl, indicate just how much time and effort OCD patients devote to *nonfunctional* actions, relative to controls. Nonfunctional, ritualized behaviors constituted more than half of OCD patients' motor acts in these tasks.[67]

Logic is not enough. Usually sooner, but nearly always later, people with OCD come to appreciate how illogical their compulsions are.[68] Often, their compulsive rituals have no connection whatsoever with their putative goal.[69] Most people with OCD are keenly aware of the uselessness of their behaviors, yet, as Ciarrocchi stresses with regard to scrupulosity,[70] attempting to argue people out of their cycles of obsessions and compulsions is almost always fruitless. Their problem is not exclusively intellectual or even cognitive. It also has a significant emotional dimension.

Whether they constitute rationally justifiable measures or not, compulsions are reinforcing, since they (temporarily) relieve the anxiety the obsession provokes. Parents check on sleeping infants to make sure that they are breathing. The act of checking produces the sight of the infant, who is breathing and well. It reassures anxious parents and relieves their distress. That relief, however, is only *temporary*.[71] The relief from the checking is good only as long as the infant is in sight. Returning to other tasks permits the anxiety to recur. That *never* ceases to be true.

But things are even worse; this cycle not only permits the return of the anxiety but also *encourages* it. That carrying out the compulsive act brings relief in the short term to the anxiety that the obsession occasions vindicates the existence of the anxiety. The relief that the compulsive act produces seduces patients into inferring that the anxiety was warranted and merited the checking. It thereby reinforces the obsession in the longer haul, which prolongs the time for the cement around the obsessive-compulsive cycle to cure and harden.[72]

Theorizing about Underlying Cognitive Mechanisms

The tremendous variability of obsessions and compulsions poses formidable theoretical challenges concerning the mechanisms that figure in OCD. Progress depends on finding stable patterns underlying that diversity. Across cultures, six traits, which we have already at least touched on, appear to inform the innumerable pairs of obsessions and compulsions that OCD patients manifest.[73] They are:

(1) An inflated sense of personal responsibility
(2) Attributing undue importance to (the occurrence of) thoughts (and their contents)
(3) Excessive concern with controlling thoughts
(4) Overestimation of hazards
(5) Intolerance of uncertainty
(6) (Misplaced) perfectionism

Cognitive models about aspects of OCD, in addition to bridging the phenomenology and neuropsychology of OCD, open a further avenue of inquiry. Since they appeal to the operations of cognitive mechanisms that *all* humans possess, they situate the features of OCD on a continuum with the mental lives and conduct of the general population.[74] The nonclinical population appears to experience uninvited negative thoughts and impulses far more regularly than generally assumed. Periodically experiencing intrusive thoughts about dangers, which people either find only temporarily troublesome or dismiss immediately, probably constitutes an *intermediate* position on this continuum. Remember that 90 percent of human beings report experiencing intrusive thoughts.[75] In a summary of papers examining hundreds of participants from thirteen countries, David Clark and Adam Radomsky note that majorities of respondents in all thirteen disclosed that they spontaneously experienced such unpleasant, unbidden thoughts.[76] Experimentally induced stress on participants who have not been diagnosed with OCD increases both the frequency and the troublesomeness of such intrusive thoughts.[77] Under normal circumstances, though, nonclinical populations face such thoughts much less often than people with OCD. They also experience far less psychological turmoil and find such thoughts easier to reject.[78]

Pierre Liénard and Pascal Boyer emphasize that at particular points in human life history, experiencing such unwanted thoughts is the *rule* rather than the exception. Most parents-to-be and parents of infants, especially, experience the same kinds of unsolicited, unwelcome thoughts. In their cases, though, these thoughts tend to center overwhelmingly on "unacceptable ideas, thoughts, urges, and images about infants."[79] This sort of preoccupation with the welfare of offspring is *common* among parents of children at this most vulnerable stage of life.

Security Motivation Systems

OCD is estimated to vex 1–3 percent of human beings at some point in their lives.[80] That range seems to hold across cultures.[81] That base rate substantially surpasses standard estimates of the mutation rate,[82] which indicates that this disorder may have connections with adaptive traits that inform human psychology and that have resulted from natural (or, perhaps, cultural) selection. We have just noted one finding that points in that direction. Even for the general population, obsessions tend to arise at points in life that are critical to successful reproduction.

Henry Szechtman and Erik Woody advance a security motivation theory of OCD that pinpoints how the system misfires.[83] They cite various evolved, automatic, domain-specific psychological systems associated with several sources of jeopardy to human survival and reproduction. They describe the operations of these psychological systems as being fast and largely isolated informationally (i.e., information in other parts of the cognitive system does not usually affect their operations).

One argument for their evolutionary roots that advocates repeatedly advance picks up on threats that humans find automatically absorbing.[84] Humans find predators, contaminants, contagious diseases, social affronts, and harm to offspring transfixing. By contrast, many far more dangerous modern perils, such as cars, cigarettes, air pollution, and electrical hazards, often leave people completely unmoved.

These security systems are concerned with *evolutionarily significant* dangers. These security motivation systems should be vigilant and very fast, and they should not be terribly discriminating. Their operations should not depend on reflective consultation of memories or inductions from previous encounters with the danger in question. Any errors they make should be false positives (which are *far* less costly than false negatives). These security-related systems should be keenly sensitive to what are often oblique and subtle indications of potential peril—a motion of the bushes, an unfamiliar odor, an unexpected sound from behind. Szechtman and Woody comment that these systems should be easily triggered but should deactivate only gradually, possessing an "easy-to-turn-on, hard-to-turn-off quality."[85]

Turning these systems *on* means that organisms are poised *to act in quite specific ways*, depending on the threat—whether to surveil the surroundings, become inconspicuous, flee predators, guard resources, avoid contaminants, or evade hazardous organisms. It is the *specificity* not only of those automatic precautionary actions but also of the patterns of learning, the processing principles, and the particular impairments associated with these systems that Boyer and Liénard seize on as evidence of those systems' relative autonomy and domain specificity.[86] Experimental evidence pertaining to precautionary rules reveals that such matters elicit the same remarkably elevated performance with conditional inference that other evolutionarily salient topics do.[87]

None of this is dysfunctional. Attention to and rapid response to such dangers, on balance, are good things. They enhance fitness. At least two problems, though, arise with OCD. The first concerns a misappraisal of circumstances. Patients devote a great deal of time, attention, and resources to matters that do not merit such outlays. The second problem is *how to turn off these systems*. Both problems are consistent with OCD patients' sense that things are somehow always wanting and that tasks and arrangements are forever incomplete, provoking compulsions to do more.[88]

A combination of four considerations, according to Szechtman and Woody, are responsible for things constantly feeling amiss. First, OCD patients feel concern because they, like everyone sooner or later, face circumstances (legitimate dangers or not) that activate their threat-specific psychological systems pertaining to predation, contamination, access to resources, and other threats. Szechtman and Woody's second factor is grounded on a logical point. Nothing in the external environment can supply airtight assurances against these potential pitfalls. Because it is logically impossible to definitively establish the absence of dangers on the basis of environmental inputs, *the termination of these systems' processing must be based on something other than environmental stimuli.*[89] With no possible environmental input logically sufficient to justify shutting down the systems' operations (checking the stove does not change the stove), they reason that "goal completion is normally signaled by an endogenously generated terminator (experienced as a feeling of knowing)" that should arise on the basis of carrying out the precautionary behavior.[90] That *internal* device that shuts down these systems' operations is their third factor. Under normal operating procedures, it shuts the system down upon completion of the precautionary behavior (*not* on the basis of any change in the environment). The fourth and pivotal factor is that with OCD patients something about that device is faulty, such that they are unable to obtain closure; they are unable to generate, experience or, perhaps, recall that *feeling of knowing.*[91]

Thus, in Szechtman and Woody's theory, when, in the presence of *perceived* dangers, evolved security systems alert people with OCD, those people are unable to resolve the attendant anxieties. They cannot relieve the distress about the perceived hazard, because *they do not experience the feeling of knowing* that adequate precautions are in place, that such precautions have been taken, or that compelling evidence indicates that the threat is no longer plausible. Ultimately, in Szechtman and Woody's view,[92] the problem is about the absence of an emotional signal.

This illustrates these systems' informational encapsulation. Patients with OCD may *consciously understand* that reasonable measures have been taken to nullify a potential threat, but *they do not experience the characteristic feeling of knowing* that should accompany that internal device's termination of operation.[93]

Religions Encourage Obsessions and Compulsions

Striking Resemblances

The behavioral and phenomenological profiles associated with OCD include abundant representation of beliefs and practices that mirror many religious beliefs and practices. Correspondingly, features of many religious beliefs and practices coincide with those behavioral and phenomenological profiles of OCD.

Their *behavioral* similarities are sometimes striking. That many religious activities closely resemble OCD patients' compulsive behaviors is uncontroversial. Consider the morning hand-washing ritual of Orthodox Jews:

> In the morning, one should wash their hands in the following manner. Before washing, one should avoid touching their mouth, nose, eyes or ears, and wash the hands as follows (for individuals who are right-handed): (1) Pick up the cup with the right hand and fill it with water. (2) Pass it to the left. (3) Pour water over the right hand and pass the cup to the left. (4) Alternate back and forth to wash three times on each hand. (5) The blessing over washing hands is then recited and the impure water should be discarded immediately. If the impure water is touched, it may require rewashing.[94]

This ritual is no less rigid and nearly as repetitive, and many parts of it are just as unnecessary as the OCD patient's ritual for turning on her television.

Decontaminating, cleansing, bathing, and purifying things and people by some means or other have been standard fare in the ritual systems of all the largest religions and of countless smaller religions throughout history. Religious ritual is not the only human activity that is as remote from rationally guided instrumental action as the compulsions of OCD patients, but neither games, nor sports, nor the arts turn so unfailingly as religious rituals do on engaging the same security motivation systems implicated in OCD.

Religions have so reliably evolved to include rituals[95] because, by signaling the need to deal with possible dangers, rituals are practices that are capable of *arresting human attention*. Boyer and Liénard describe this as their capacity to achieve "cognitive capture" of humans' security motivation systems.[96] Cultural representations mimic some cue from the collection of stimuli that activate the system. Crucially, these maturationally natural sensitivities and the associated precautionary imperatives and default inferences are *automatic and beyond conscious control*.[97] Religious representations' engagement of humans' security motivation systems makes rituals "attention-grabbing, intuitively appropriate, and compelling."[98] These are just the properties that make religious representations cognitively enticing and give them a *transmission advantage* in the de facto competition to communicate and spread cultural representations in populations of human minds.[99] The conception of those rituals as appropriate precautionary behaviors and the careful specification of their details endow them with an unavoidable psychological salience and urgency.[100]

Religious rituals that piggyback on the sensitivities of humans' security motivation systems activate the psychological process that Boyer and Liénard call "ritualization." Ritualization results in parsing actions at uncharacteristically low levels, overwhelming working memory, demoting goals, and thoroughly distracting conscious cognition, all of which collectively render participants thoroughly receptive to religious authorities' accounts of what has transpired.[101]

The *phenomenological* profiles of OCD and of many aspects of religious life are similar, too. One prominent feature is the emphasis that many of the most successful religions place on *explicit* versions of moral TAF. Boyer argues that conceiving of gods as persons is the most salient feature of religions and that humans' cognitive systems for managing in the social world automatically represent "the extent to which other people have strategic information" and what strategic information they have.[102] Strategic information is any information available to an agent about some situation that "*activates the mental systems that regulate social interaction*," such as theory of mind.[103] What counts as strategic information can change, since whether something qualifies depends on the social context and its impact on a person's social inference systems. As Boyer observes, "If the literary preferences of your dinner companion have no special effect on the ongoing interaction, then it remains plain information; but it becomes strategic information if it triggers inferences about what to do next."[104] The key point is that an overwhelming majority of the most successful religions have gods who have *full access* to strategic information. These gods know not only what we do but also what we think. They know every piece of strategic information there is. Since these gods are also the putative authors of human morality, they are morally interested parties. They know every morally relevant detail about human beings' actions and *about their states of mind* informing those actions. The gods also police them. Human beings are every bit as responsible for their sinful thoughts as for their sinful actions. They deserve divine punishments no less for the former than for the latter, and one is no less obvious than the other *to gods who have full access to strategic information*.[105]

Morally interested, thought-monitoring gods set the stage for religiously induced moral TAF, but in the event that any ambiguity remains, many religions explicitly affirm it. The proscription against covetousness, for example, is the prohibition of *a state of mind*.

Can Religion Cause Scrupulosity and Domesticate OCD in the Bargain?

This overlap between features of religions and features of OCD behaviorally and phenomenologically convinces researchers about their connections, but clarifying what those connections might be is complicated. One question is whether religions and religiously inspired moral TAF, in particular, can *cause* scrupulosity, construed as the religious expression of OCD. *That* question, however, is ambiguous. We take up two possible interpretations—both endorsed by Abramowitz and Jacoby. Both also count as species of what we have referred to as the "domestication" of OCD expressed as scrupulosity.[106]

Can religious arrangements and, in particular, religiously inspired moral TAF

(1) maintain and exacerbate existing scrupulosity?

(2) increase the rate at which OCD is expressed as scrupulosity?

Testimonies by the likes of Bunyan and Luther offer striking evidence for responding affirmatively to interpretation (1)—making for the first species of domestication. Their cases make sense of asking whether religious representations can incite misinterpretations of wayward intrusive thoughts, thereby maintaining or even exacerbating scrupulosity where it already exists in people with OCD. What their works and lives show is that some religious arrangements serve to channel, maintain, and probably exacerbate preexisting cases of scrupulosity.

Especially when it is explicitly affirmed by the ruler(s) of the universe, religiously inspired moral TAF has at least *two important psychological corollaries* that echo familiar patterns with OCD. When peoples' fates for eternity might turn on the moral and religious statuses of their thoughts, religiously inspired moral TAF undoubtedly reinforces any existing convictions and probably instills newfound convictions, first, about those thoughts' significance and, second, about the importance of their control.

Abramowitz and Jacoby hold that such religiously inspired moral TAF and these corollaries incite misinterpretations of wayward intrusive thoughts that *maintain and probably exacerbate scrupulosity where it already exists.*[107] Multiple studies indicate that OCD symptoms are amplified in religiously scrupulous individuals. Compared to other OCD patients, these people manifest less insight about the irrationality of their obsessions and more perceptual distortions and magical ideation, such as likelihood TAF.[108] They also have an elevated sense of the significance of their thoughts, their need to control their thoughts, and their personal responsibility. Those effects, in turn, seem to lead to comparatively greater perfectionism in this population as well.[109] The evidence overall seems in line with Abramowitz and Jacoby's suggestion. It points to aggravated versions of familiar patterns.[110] In such religious contexts, the scrupulous might go unnoticed, blending in with mainstream participants.[111] Distinguishing the scrupulous from the religiously dedicated might prove challenging at best, especially in the short run.

That religious manifestations of OCD will turn on the cultural context and the distinctive contents of the local religion is not too controversial, but interpretation (2) of the causal question considers a stronger version of domestication. It asks whether some religious arrangements and religiously inspired moral TAF, in particular, can increase the rate of scrupulosity within a culture *among people with OCD*. Or, more generally, can some religions increase the rate among those with OCD at which that disorder gets expressed as scrupulosity?

Abramowitz and Jacoby propose that religiously inspired misreadings of everyday intrusions increase the probability that intrusive thoughts "may develop into clinical obsessions."[112] They suggest, in effect, that religiously inspired moral TAF is likely to *increase the rate at which OCD receives religious expression as scrupulosity in a culture*.

Assessing this possibility is abetted by cross-cultural comparisons since, presumably, some religions may be more likely than others to generate such responses among those with OCD. Although comparisons across cultures suggest that on many fronts—including everything from the topics of intrusive obsessions, to the major categories of compulsions, to the core traits listed earlier—OCD varies little,[113] studies of different cultures and religions suggest that they *do* substantially influence whether OCD symptoms are manifested religiously.[114] The rates of scrupulosity among people with OCD vary considerably between religions and between cultures. Nearly a quarter of OCD patients in one large American study showed *some* religious obsessions, and nearly 6 percent of American patients in an even larger study showed religious obsessions *primarily*.[115] By contrast, studies in Egypt and Saudi Arabia have found rates of religious scrupulosity (i.e., OCD patients showing religious obsessions primarily) as high as *sixty* percent among their patient populations.[116] The greater prominence of religions endorsing moral TAF in some cultures results in higher rates *of manifestations of OCD as scrupulosity* in those cultures.[117]

These two forms of domestication, though, fall short of Cefalu's claim for religious arrangements capable of yielding a "non-pathological obsessive religiosity."[118] We have already noted, first, that such arrangements may mainstream scrupulosity, but they do not cure it. Here we briefly scout a second counterargument. Religions often pose barriers in the treatment of scrupulous patients. The simple availability of clergy or other religious leaders can serve to intensify OCD in religious followers.[119] These people can be sources of reassurance when responding to compulsive consultations. Unbeknownst to sympathetic clergy (initially at least), their reassurances provide the temporary relief that reinforces the scrupulous individual's compulsive consultations. (Both Luther and Loyola would confess the same sinful incidents repeatedly.) Ironically, clergy's misplaced support can constitute a means for "recovery avoidance" in their scrupulous followers.[120] Huppert and his colleagues outline the importance of enlisting the rabbis of scrupulous Orthodox Jews in the administration of their treatment. Getting the rabbis to *refuse* to answer their scrupulous followers' questions is a way to signal to those followers that "they are dealing with OCD and not true religious issues."[121]

To co-opt a point that Flanagan makes about the disorder of addiction, religiously induced scrupulosity is, in effect, an *ironic* disorder[122]—religions serve as both the causes and the cures, as both the arsonists and the fire brigades[123]—again and again. Religions lay out the norms for full membership in the communities that they define. That *always* includes participating in the ritual lives of those communities, and for a host of reasons, *the rituals never stop*. That is because, first, religious ritual systems encompass all of life. As Flanagan notes, "the space of spirituality and religion [seems] designed to function comprehensively for those for whom it is designed to appeal—in principle to everyone in the vicinity."[124] As part of that comprehensive functioning, religions use rituals to mark significant events in the human life course, such as births, marriages, and deaths. Second, participants must do some rituals repeatedly. No amount of ritual performance will ever satisfy the appetites of the gods. In addition, religious ritual systems are self-reinforcing. So, for example, perpetuating religious communities depends on iterations of rituals (e.g., initiations), the performances of which presume successful performances of more rituals (e.g., certifying ritual practitioners).[125] Some further ritual *always* awaits accomplishment. With the gods' seal of approval from above, the perpetual temptation to be a ritualistic overachiever never goes away. For both the faithful and the scrupulous, the impetus is compelling.

Notes

1. James, The sentiment of rationality, 372.

2. See Flanagan, *Dreaming Souls.*

3. See McCauley and Graham, *Hearing Voices and Other Matters of the Mind.*

4. Flanagan, *Dreaming Souls,* 14.

5. Bunyan, *Grace Abounding to the Chief of Sinners.*

6. Ibid., 8.

7. Ibid., 11.

8. Ibid., 36.

9. American Psychiatric Association, *Diagnostic and Statistical Manual of Mental Disorders, Fifth Edition.*

10. Excessive conscientiousness is a trait more directly associated, perhaps, with obsessive compulsive personality disorder (OCPD), a less severe, less persistent, and less troubling condition that otherwise resembles OCD.

11. Wortman, *Triggered,* 2.

12. Abramowitz and Jacoby, Scrupulosity, 147.

13. Nelson et al., Scrupulosity in patients with obsessive-compulsive disorder, 1072, 1082–1083.

14. Summers and Sinnott-Armstrong, Scrupulous agents.

15. Ibid., 950–951; Abramowitz and Jacoby, Scrupulosity.

16. Abramowitz and Jacoby, Scrupulosity, 145. Arguably, these apparently hyper-religious individuals are *less* religious than their nonscrupulous brethren!

17. Greenberg, Are religious compulsions religious or compulsive?; Ciarrocchi, *The Doubting Disease.*

18. Bunyan, *Grace Abounding to the Chief of Sinners,* 27, 64; Cefalu, The doubting disease, 117; Ciarrocchi, *The Doubting Disease,* 36.

19. Bunyan, *Grace Abounding to the Chief of Sinners,* 36.

20. Abramowitz and Jacoby, Scrupulosity, 144; Nelson et al., Scrupulosity in patients with obsessive-compulsive disorder, 1072.

21. Bunyan, *Grace Abounding to the Chief of Sinners*, 53.

22. Ibid., 19.

23. Ibid., 43–44.

24. Ibid., 53. Bunyan is not unique. Martin Luther describes many of the same symptoms in his autobiographical writings.

25. Flannelly, *Religious Beliefs, Evolutionary Psychiatry, and Mental Health in America*, 245–246.

26. Cefalu, The doubting disease, 114.

27. Shefran, Thordarson, and Rachman, Thought-action fusion in obsessive compulsive disorder, 379.

28. Cefalu, The doubting disease, 115.

29. See, for example, Preus, *Sola Gratia*.

30. Cefalu, The doubting disease, 117. How relevant such arrangements may be for nonscrupulous Protestants, the vast majority of Protestants, would seem a different matter.

31. For example, Bunyan, *Grace Abounding to the Chief of Sinners*, 24.

32. Cefalu, The doubting disease, 119.

33. Fiske and Haslam, Is obsessive-compulsive disorder a pathology of the human disposition to perform socially meaningful rituals?

34. Luhrmann, *When God Talks Back*.

35. Dulaney and Fiske, Cultural rituals and obsessive-compulsive disorder, 249.

36. Cefalu, The doubting disease, 124.

37. Ciarrocchi, *The Doubting Disease*, 123.

38. Rassin and Koster, The correlation between thought–action fusion and religiosity in a normal sample; Abramowitz et al., Association between Protestant religiosity and obsessive-compulsive symptoms and cognitions, 74.

39. Abramowitz et al., Association between Protestant religiosity and obsessive-compulsive symptoms and cognitions.

40. Abramowitz and Jacoby, Scrupulosity, 141, 144.

41. Ciarrocchi, *The Doubting Disease*, 48–49.

42. Yorulmaz, Gencoz, and Woody, OCD cognitions and symptoms in different religious contexts, 403–404.

43. Abramowitz and Jacoby, Scrupulosity, 142; see also Zohar et al., Religiosity and obsessive-compulsive behavior in Israeli Jews.

44. Abramowitz and Jacoby, Scrupulosity, 144.

45. Rapoport and Fiske, The new biology of obsessive-compulsive disorder, 168.

46. Huppert, Siev, and Kushner, When religion and obsessive-compulsive disorder collide, 926.

47. Ciarrocchi, *The Doubting Disease*, 21; see also Singer and Gillihan, *Overcoming OCD*. A recent attempt to order this heterogeneous collection of symptoms used exploratory factor analysis and Bayesian structural equation modeling to ascertain underlying factors that inform the range of particular symptoms connected with OCD from one sample of patients, which were cross-validated with a second sample. This analysis arrived at ten first-order factors:

Keeping order	Pure repetition
Magical thinking	Mental urges
Mental exactness	Sexual obsessions
Aggressive impulses	Responsibility
Somatic obsessions	Cleanliness

which, upon further analysis, reduced to four second-order factors: contamination, taboo thoughts, responsibility, and incompleteness (where the latter factor includes diverse miscellaneous symptoms). See Schulze, Kathmann, and Reuter, Getting it just right, 63–73.

48. Eilam et al., Rituals, stereotypy and compulsive behavior in animals and humans, 469.

49. Seligman, Walker, and Rosenham, *Abnormal Psychology*, 207; Ruscio et al., The epidemiology of obsessive-compulsive disorder in the National Comorbidity Survey Replication, 58.

50. American Psychiatric Association, *Diagnostic and Statistical Manual of Mental Disorders, Fifth Edition*.

51. Salkovskis, Obsessional-compulsive problems; Rachman and Shafran, Cognitive behavioural features of obsessive-compulsive disorder.

52. Summers and Sinnott-Armstrong, Scrupulous agents, 949.

53. Culture can play a role here, though. See, for example, Okasha et al., Phenomenology of obsessive-compulsive disorder.

54. Wortman, *Triggered*, 198–199.

55. Ibid., 31–32.

56. Summers and Sinnott-Armstrong, Scrupulous agents, 958.

57. Ibid.

58. Brakoulias and Starcevic, The characterization of beliefs in obsessive-compulsive disorder.

59. Eilam et al., Rituals, stereotypy and compulsive behavior in animals and humans, 469.

60. Singer and Gillihan, *Overcoming OCD*, 40.

61. Wegner et al., Paradoxical effects of thought suppression.

62. Ciarrocchi, *The Doubting Disease*, 160.

63. Szechtman and Woody, Obsessive-compulsive disorder as a disturbance of security motivation, 112.

64. Ruscio et al., The epidemiology of obsessive-compulsive disorder in the National Comorbidity Survey Replication, 56.

65. Rachman, A cognitive theory of compulsive checking, 627.

66. Polimeni, Reiss, and Sareen, Could obsessive-compulsive disorder have originated as a group-selected adaptive trait in traditional societies?, 659.

67. Zor et al., Obsessive-compulsive disorder.

68. Feygin, Swain, and Leckman, The normalcy of neurosis, 855; Flannelly, *Religious Beliefs, Evolutionary Psychiatry, and Mental Health in America*, 97.

69. Nielbo and Sørensen, Spontaneous processing of functional and non-functional action sequences.

70. Ciarrocchi, *The Doubting Disease*, 113.

71. Seligman, Walker, and Rosenham, *Abnormal Psychology*, 212.

72. See Summer and Sinnott-Armstrong, Scrupulous agents, 960.

73. Nelson et al., Scrupulosity in patients with obsessive-compulsive disorder, 1073; Fontenelle et al., Trans-cultural aspects of obsessive-compulsive disorder.

74. Sica, Navara, and Sanavio, Religiousness and obsessive-compulsive cognitions and symptoms in an Italian population, 815; Zor et al., Obsessive-compulsive disorder, 296.

75. Feygin, Swain, and Leckman, The normalcy of neurosis, 856.

76. Clark and Radomsky, Introduction.

77. Seligman, Walker, and Rosenham, *Abnormal Psychology*, 211.

78. Ciarrocchi, *The Doubting Disease*, 25; Feygin, Swain, and Leckman, The normalcy of neurosis, 856.

79. Boyer and Liénard, Why ritualized behavior?, 619.

80. Zor et al., Obsessive-compulsive disorder, 288.

81. Polimeni, Reiss, and Sareen, Could obsessive-compulsive disorder have originated as a group-selected adaptive trait in traditional societies?, 656.

82. Ibid.

83. Ibid., 114; Woody and Szechtman, Adaptation to potential threat.

84. For example, see Boyer, *Minds Make Societies.*

85. Szechtman and Woody, Obsessive-compulsive disorder as a disturbance of security motivation, 114.

86. Boyer and Liénard, Why ritualized behavior?, 602–603.

87. Fiddick, Cosmides, and Tooby, No interpretation without representation.

88. Zor et al., Obsessive-compulsive disorder, 295–296.

89. Woody and Szechtman, Uncertainty and rituals, 634.

90. Szechtman and Woody, Obsessive-compulsive disorder as a disturbance of security motivation, 116.

91. Ibid., 113.

92. Ibid., 123.

93. Ibid., 122.

94. Huppert, Siev, and Kushner, When religion and obsessive-compulsive disorder collide, 930. For an illustration from Catholicism bearing on the same point, see Wortman, *Triggered*, 35–36.

95. Exceptions are few—the Society of Friends (i.e., the Quakers) might qualify as a *partial* exception.

96. Boyer and Liénard, Why ritualized behavior?, 609–610.

97. McCauley, *Why Religion Is Natural and Science Is Not.*

98. Boyer and Liénard, Why collective rituals?, 814; Boyer and Liénard, Why ritualized behavior?, 611, 640.

99. Boyer and Liénard, Why collective rituals?, 823.

100. Boyer and Liénard, Why ritualized behavior?, 609.

101. Schjoedt et al., Cognitive resource depletion in religious interactions.

102. Boyer, *Religion Explained*, 154.

103. Ibid., 152.

104. Ibid.

105. Ara Norenzayan's Big Gods theory and its intellectual progeny touch on related topics, but that theory focuses on the gods' monitoring of humans' *behaviors*. See Norenzayan, *Big Gods*. We suggest elsewhere (McCauley and Graham, *Hearing Voices and Other Matters of the Mind*) that religiously induced moral TAF and the moral monitoring of *thoughts* can also play a significant role in the persistence of religions.

106. Such domestication of OCD is to be contrasted with what we call *enlistment* (for which we argue in McCauley and Graham, *Hearing Voices and Other Matters of the Mind*). Enlistment is the thesis that religious arrangements elicit features of scrupulosity directly, though only temporarily, in the *general* population.

107. Abramowitz and Jacoby, Scrupulosity, 140–149. Other researchers concur. See, for example, Ciarrocchi, *The Doubting Disease*, 123–124; Steketee, Quay, and White, Religion and guilt in OCD patients, 360.

108. Abramowitz and Jacoby, Scrupulosity; Nelson et al., Scrupulosity in patients with obsessive-compulsive disorder.

109. Yorulmaz, Gencoz, and Woody, OCD cognitions and symptoms in different religious contexts.

110. Some studies of OCD and religion in some cultures have not yielded evidence associating levels of religiosity and religious obsessions. See, for example, Tek and Ulug, Religiosity and religious obsessions in obsessive-compulsive disorder.

111. Abramowitz and Jacoby, Scrupulosity; Nelson et al., Scrupulosity in patients with obsessive-compulsive disorder, 141.

112. Nelson et al., Scrupulosity in patients with obsessive-compulsive disorder, 142.

113. Rapoport and Fiske, The new biology of obsessive-compulsive disorder; Fontenelle et al., Trans-cultural aspects of obsessive-compulsive disorder; Huppert, Siev, and Kushner, When religion and obsessive-compulsive disorder collide.

114. Abramowitz et al., Association between Protestant religiosity and obsessive compulsive symptoms and cognitions, 71; Tek and Ulug, Religiosity and religious obsessions in obsessive-compulsive disorder, 106; Steketee, Quay, and White, Religion and guilt in OCD patients, 360; Yorulmaz, Gencoz, and Woody, OCD cognitions and symptoms in different religious contexts, 401–402; Zohar et al., Religiosity and obsessive-compulsive behavior in Israeli Jews, 858–859.

115. Nelson et al., Scrupulosity in patients with obsessive-compulsive disorder.

116. Okasha et al., Phenomenology of obsessive-compulsive disorder; Abramowitz and Jacoby, Scrupulosity.

117. Greenberg and Shefler, Obsessive compulsive disorder in ultra-orthodox Jewish patients; Yorulmaz, Gencoz, and Woody, OCD cognitions and symptoms in different religious contexts.

118. Cefalu, The doubting disease, 117.

119. Abramowitz and Jacoby, Scrupulosity, 146.

120. Singer and Gillihan, *Overcoming OCD*, 157.

121. Huppert, Siev, and Kushner, When religion and obsessive-compulsive disorder collide, 937. They report that, frustrated from endless appointments with scrupulous members of their communities, the rabbis usually welcome this approach. (See also Ciarrocchi, *The Doubting Disease*, 76–77.)

122. Flanagan, Identity and addiction, 77.

123. I borrow a metaphor from Norenzayan, *Big Gods*.

124. Flanagan, *The Really Hard Problem*, 189.

125. Lawson and McCauley, *Rethinking Religion*; McCauley and Lawson, *Bringing Ritual to Mind*.

Bibliography

Abramowitz, J. S., Deacon, B. J., Woods, C. M., and Tolin, D. F. Association between Protestant religiosity and obsessive-compulsive symptoms and cognitions. *Depression and Anxiety* 20 (2004): 70–76.

Abramowitz, Jonathan S., and Jacoby, Ryan J. Scrupulosity: A cognitive-behavioral analysis and implications for treatment. *Journal of Obsessive-Compulsive and Related Disorders* 3 (2014): 140–149.

American Psychiatric Association. *Diagnostic and Statistical Manual of Mental Disorders, Fifth Edition: DSM-5*. Washington, DC: American Psychiatric Association, 2013.

Boyer, Pascal. *Minds Make Societies: How Cognition Explains the World Humans Create*. New Haven, CT: Yale University Press, 2018.

Boyer, Pascal. *Religion Explained: The Evolutionary Origins of Religious Thought*. New York: Basic Books, 2001.

Boyer, Pascal, and Liénard, Pierre. Why collective rituals? A cultural selection model of ritualized behavior. *American Anthropologist* 108, no. 4 (December 2006): 814–828.

Boyer, Pascal, and Liénard, Pierre. Why ritualized behavior? Precaution systems and action parsing in developmental, pathological and cultural rituals. *Behavioral and Brain Sciences* 29, no. 6 (December 2006): 595–650.

Brakoulias, V., and Starcevic, V. The characterization of beliefs in obsessive-compulsive disorder. *Psychiatric Quarterly* 82, no. 2 (June 2010): 151–161.

Bunyan, John. *Grace Abounding to the Chief of Sinners*. 1666. Repr., London: Penguin, 1987.

Cefalu, Paul. The doubting disease: Religious scrupulosity and obsessive-compulsive disorder in historical context. *Journal of Medical Humanities* 31, no. 2 (June 2010): 111–125.

Ciarrocchi, J. W. *The Doubting Disease: Help for Scrupulosity and Religious Compulsions*. Mahwah, NJ: Paulist Press, 1995.

Clark, D. A., and Radomsky, A. S. Introduction: A global perspective on unwanted intrusive thoughts. *Journal of Obsessive-Compulsive and Related Disorders* 3, no. 3 (July 2014): 265–268.

Dulaney, S., and Fiske, A. P. Cultural rituals and obsessive-compulsive disorder: Is there a common psychological mechanism? *Ethos* 22, no. 3 (September 1994): 243–283.

Eilam, D., Zor, R., Szechtman, H., and Hermesh, H. Rituals, stereotypy and compulsive behavior in animals and humans. *Neuroscience and Behavioral Reviews* 30, no. 4 (February 2006): 456–471.

Feygin, D. L., Swain, J. E., and Leckman, J. F. The normalcy of neurosis: Evolutionary origins of obsessive-compulsive disorder and related behaviors. *Progress in Neuro-psychopharmacology and Biological Psychiatry* 30, no. 5 (July 2006): 854–864.

Fiddick, L., Cosmides, L., and Tooby, J. No interpretation without representation: The role of domain-specific representations and inferences in the Wason selection task. *Cognition* 77, no. 1 (October 2000): 1–79.

Fiske, A. P., and Haslam, N. Is obsessive-compulsive disorder a pathology of the human disposition to perform socially meaningful rituals? Evidence of similar content. *Journal of Nervous and Mental Disease* 185, no. 4 (April 1997): 211–222.

Flanagan, Owen. *Dreaming Souls: Sleep, Dreams, and the Evolution of the Conscious Mind*. New York: Oxford University Press, 2000.

Flanagan, Owen. Identity and addiction. In *The Routledge Handbook of Philosophy and Science of Addiction*, edited by H. Pickard and S. Ahmed, 77–89. London: Routledge, 2019.

Flanagan, Owen. *The Really Hard Problem: Meaning in a Material World*. Cambridge, MA: MIT Press, 2007.

Flannelly, Kevin J. *Religious Beliefs, Evolutionary Psychiatry, and Mental Health in America: Evolutionary Threat Assessment Systems Theory*. Cham, Switzerland: Springer, 2017.

Fontenelle, L. F., Mendlowicz, M. V., Marques, C., and Versiani, M. Trans-cultural aspects of obsessive-compulsive disorder: A description of a Brazilian sample and a systematic review of international clinical studies. *Journal of Psychiatric Research* 38, no. 4 (July–August 2004): 403–411.

Greenberg, D. Are religious compulsions religious or compulsive? *American Journal of Psychotherapy* 38, no. 4 (October 1984): 524–532.

Greenberg, D., and Shefler, G. Obsessive compulsive disorder in ultra-orthodox Jewish patients: A comparison of religious and non-religious symptoms. *Psychology and Psychotherapy: Theory, Research and Practice* 75, no. 2 (June 2002): 123–130.

Huppert, J. D., Siev, J., and Kushner, E. S. When religion and obsessive-compulsive disorder collide: Treating scrupulosity in ultra-Orthodox Jews. *Journal of Clinical Psychology* 63, no. 10 (October 2007): 925–941.

James, William. The sentiment of rationality. In *The Many Faces of Philosophy: From Plato to Arendt*, edited by A. Rorty, 370–373. Oxford: Oxford University Press, 2003.

Lawson, E. T., and McCauley, R. N. *Rethinking Religion: Connecting Cognition and Culture*. Cambridge: Cambridge University Press, 1990.

Luhrmann, T. *When God Talks Back: Understanding the American Evangelical Relationship with God*. New York: Knopf, 2012.

McCauley, Robert. *Why Religion Is Natural and Science Is Not*. New York: Oxford University Press, 2011.

McCauley, Robert, and Graham, George. *Hearing Voices and Other Matters of the Mind: What Mental Abnormalities Can Teach Us about Religions*. New York: Oxford University Press, 2020.

McCauley, Robert, and Lawson, E. T. *Bringing Ritual to Mind: Psychological Foundations of Cultural Forms*. Cambridge: Cambridge University Press, 2002.

Nelson, E. A., Abramowitz, J. S., Whiteside, S. P., and Deacon, B. J. Scrupulosity in patients with obsessive-compulsive disorder: Relationship to clinical and cognitive phenomena. *Journal of Anxiety Disorders* 20, no. 8 (March 2006): 1071–1086.

Nielbo, Kristoffer L., and Sørensen, Jesper. Spontaneous processing of functional and non-functional action sequences. *Religion, Brain and Behavior* 1, no. 1 (February 2011): 18–30.

Norenzayan, Ara. *Big Gods: How Religion Transformed Cooperation and Conflict*. Princeton, NJ: Princeton University Press, 2013.

Okasha, A., Saad, A., Khalil, A. H., Seif El Dawla, A., and Yehia, N. Phenomenology of obsessive-compulsive disorder: A transcultural study. *Comparative Psychiatry* 35, no. 3 (May–June 1994): 191–197.

Polimeni, J., Reiss, J. P., and Sareen, J. Could obsessive-compulsive disorder have originated as a group-selected adaptive trait in traditional societies? *Medical Hypotheses* 65, no. 4 (2005): 655–664.

Preus, Stephen. *Sola Gratia*. LutheranReformation.org, June 13, 2006. https://lutheranreformation.org/theology/sola-gratia/.

Rachman, S. A cognitive theory of compulsive checking. *Behaviour and Research Therapy* 40, no. 6 (June 2002): 625–639.

Rachman, S., and Shafran, R. Cognitive behavioural features of obsessive-compulsive disorder. In *Obsessive-Compulsive Disorder: Theory, Research and Treatment*, edited by R. Swinson, M. Antony, S. Rachman, and M. Richter, 51–78. New York: Guilford Press, 1998.

Rapoport, J., and Fiske, A. The new biology of obsessive-compulsive disorder: Implications for evolutionary psychology. *Perspectives in Biology and Medicine* 41, no. 2 (1998): 159–175.

Rassin, E., and Koster, E. The correlation between thought–action fusion and religiosity in a normal sample. *Behavior Research and Therapy* 41, no. 3 (2003): 361–368.

Ruscio, A. M., Stein, D. J., Chiu, W. T., and Kessler, R. C. The epidemiology of obsessive-compulsive disorder in the National Comorbidity Survey Replication. *Molecular Psychiatry* 15, no. 1 (January 2010): 53–63.

Salkovskis, P. M. Obsessional-compulsive problems: A cognitive-behavioral analysis. *Behaviour Research and Therapy* 23, no. 5 (1985): 571–583.

Schjoedt, Uffe, Sorensen, Jesper, Laigaard Nielbo, Kristoffer, Xygalatas, Dimitris, Mitkidis, Panagiotis, and Bulbulia, Joseph. Cognitive resource depletion in religious interactions. *Religion, Brain and Behavior* 3, no. 1 (January 2013): 39–55.

Schulze, Daniel, Kathmann, Norbert, and Reuter, Benedikt. Getting it just right: A reevaluation of OCD symptom dimensions integrating traditional and Bayesian approaches. *Journal of Anxiety Disorders* 56 (May 2018): 63–73.

Seligman, M. E. P., Walker, E. F., and Rosenham, D. L. *Abnormal Psychology*. New York: W. W. Norton, 2001.

Shafran, R., Thordarson, D. S., and Rachman, S. Thought-action fusion in obsessive compulsive disorder. *Journal of Anxiety Disorders* 10, no. 5 (1996): 379–391.

Sica, C., Novara, C., and Sanavio, E. Religiousness and obsessive compulsive cognitions and symptoms in an Italian population. *Behaviour Research and Therapy* 40, no. 7 (July 2002): 813–823.

Singer, J., and Gillihan, S. J. *Overcoming OCD: A Journey to Recovery*. Lanham, MD: Rowman and Littlefield, 2015.

Steketee, G., Quay, S., and White, K. Religion and guilt in OCD patients. *Journal of Anxiety Disorders* 5, no. 4 (1991): 359–367.

Summers, Jesse, and Sinnott-Armstrong, Walter. Scrupulous agents. *Philosophical Psychology* 28, no. 7 (2015): 947–966.

Szechtman, H., and Woody, E. Obsessive-compulsive disorder as a disturbance of security motivation. *Psychological Review* 111, no. 1 (January 2004): 111–127.

Tek, C., and Ulug, B. Religiosity and religious obsessions in obsessive-compulsive disorder. *Psychiatry Research* 104, no. 2 (November 2001): 99–108.

Wegner, Daniel M., Schneider, David J., Carter, Samuel R., and White, Teri L. Paradoxical effects of thought suppression. *Journal of Personality and Social Psychology* 53, no. 1 (July 1987): 5–13.

Woody, E. Z., and Szechtman, H. Adaptation to potential threat: The evolution, neurobiology, and psychopathology of the security motivation system. *Neuroscience and Biobehavioral Reviews* 35, no. 4 (March 2011): 1019–1033.

Woody, Erik Z., and Szechtman, Henry. Uncertainty and rituals. *Behavioral and Brain Sciences* 29, no. 6 (2006): 634–635.

Wortman, Fletcher. *Triggered: A Memoir of Obsessive-Compulsive Disorder*. New York: St. Martin's Press, 2012.

Yorulmaz, O., Gencoz, T., and Woody, S. OCD cognitions and symptoms in different religious contexts. *Journal of Anxiety Disorders* 23, no. 3 (April 2009): 401–406.

Zohar,. A., Goldman, E., Calamary, R., and Mashiah, M. Religiosity and obsessive-compulsive behavior in Israeli Jews. *Behaviour Research and Therapy* 43, no. 7 (2005): 857–868.

Zor, R., Keren, H., Hermesh, H., Szechtman, H., Mort, J., and Eilan, D. Obsessive-compulsive disorder: A disorder of pessimal (non-functional) motor behavior. *Acta Psychiatrica Scandanavica* 120, no. 4 (October 2009): 288–298.

3 Patients as Experience-Based Experts in Psychiatry: Insights from the Natural Method

Şerife Tekin

Introduction

A primary goal of psychiatric epistemology is to identify the properties of mental disorders that are relevant for developing effective interventions. Sources of information individuating these properties include scientific research on mental disorders (e.g., clinical drug trials), data emerging from clinical settings (e.g., case studies), and first-person reports of those suffering from mental disorders or those who witness or observe such suffering. While dominant scientific frameworks in psychiatry recognize and rely on scientific and clinical data in their efforts to advance the science of mental disorders, they are not similarly committed to directly using the first-person perspectives of patients. More specifically, individuals who suffer from and are diagnosed with mental disorders have not systematically been made part of the scientific inquiry into mental disorders. This chapter calls for this problem to be rectified in scientific psychiatry. I argue that individuals diagnosed with mental disorders are "experience-based" experts on mental disorders and, as such, must be included in the scientific research aiming to expand knowledge in psychiatry. In developing this argument, I challenge the presupposition that individuals with mental disorders lack the expertise to make meaningful contributions to psychiatric epistemology.

In the first section of the chapter I review the history of the creation and revision processes of the *Diagnostic and Statistical Manual of Mental Disorders* (*DSM*) to illustrate how patients have historically been left out of scientific inquiry in mental disorders, tracing it to a presupposition that they lack the expertise required for scientific investigation. In the second section I critically examine the definition of *expertise* deployed by the American

Psychiatric Association (APA), creator of the *DSM*, and develop an alternative definition by identifying patients as experience-based experts. My discussion is guided by Owen Flanagan's natural method and the account of expertise developed by third-wave science studies scholars. My argument for including patients in psychiatric research is primarily conceptual: I challenge the assumption that individuals with mental disorders lack expertise in the scientific investigation of mental disorders. Unfortunately, while a discussion of how to develop better scientific policies (or scientific paradigms) involving patients would be interesting, it is beyond the scope of this chapter.

The Presupposition: "Patients Are Not Experts"

Historically, patients have not been included in the scientific inquiry on mental disorders as "subjects" who produce knowledge. Their role has been constrained to serving as the "objects" of study in clinical trials or clinical case studies. This is largely because of the presupposition that patients lack the expertise required for the scientific investigation of mental disorders. The goal of this section is to illustrate the pervasiveness of this presupposition by reviewing the creation and revision processes of the *DSM*. I focus on the *DSM* because it is the primary classificatory schema used in the United States and around the world to expand knowledge on mental disorders in psychiatry, with a mandate to serve "clinical, research, and educational purposes."[1] I examine the *DSM* creators' conception of what counts as "expertise" in the scientific investigation of mental disorders. As I go on to show, in the various versions of the *DSM*, the notion of "expertise" is limited to those with clinical training in psychopathology; expertise on the basis of experience is discounted.

The primary reason for developing a classification of mental disorders in the United States was the perceived need to collect statistical information on the prevalence of mental disorders. The 1840 census, conducted by the Census Office, recorded the frequency of "idiocy/insanity" in the population, but there were many errors in this preliminary classification, as reported by the American Statistical Association to the US House of Representatives.[7] In response, the Association of Medical Superintendents of American Institutions for the Insane was formed in 1844 to ensure that a classification system for mental disorders was created by medical experts.

The Association of Medical Superintendents of American Institutions, later the American Psychiatric Association (APA), worked with the National Commission on Mental Hygiene (now Mental Health America) to develop a guide for mental hospitals called *The Statistical Manual for the Use of Institutions for the Insane*. Along with the New York Academy of Medicine, the APA also developed the psychiatric nomenclature subsection of the US general medical guide *The Standard Classified Nomenclature of Disease*.[3]

The initial taxonomy for mental disorders was further developed during World War II, when psychiatrists became heavily involved in the selection, assessment, and treatment of soldiers. For example, in 1943, a committee of psychiatrists developed a new classification scheme, Medical 203, which was published in a War Department technical bulletin. Shortly after the war, in 1949, the World Health Organization (WHO) published the sixth edition of the *International Statistical Classification of Diseases* (*ICD*). For the first time, the *ICD* included a section on mental disorders.

Despite the earlier efforts to come up with a common classification system, as these examples suggest, in the mid-twentieth century, many different systems were in use, and there was mounting concern about developing a unifying schema.[4] In response to this concern, an APA Committee on Nomenclature and Statistics developed a system of classification specifically for use in the United States to standardize the diverse and often inconsistent usage of different documents listing the criteria for mental disorders. In 1950, the APA committee undertook a review of Medical 203 by consulting the members of the APA and the Veterans Administration (VA) and created the first edition of the *Diagnostic and Statistical Manual of Mental Disorders* (*DSM-I*), published in 1952. *DSM-I* had clear statistical goals: it aimed "to make possible the gathering of data for future clarification of ideas concerning etiology, pathology, prognosis, and treatment in mental disorders."[5] It was grounded in the psychoanalytic approach: mental disorders, also called "reactions," were represented in relation to the psychodynamic causal factors underlying them. These causal factors were characterized as either a dysfunction in the brain or a general difficulty in adapting to environmental stressors resulting from unresolved sexual conflicts of childhood. For example, psychotic disorders were defined as disorders of psychogenic origin or without clearly defined physical cause or structural change in the brain.[6] The following paragraph describes *who* was involved in the development of *DSM-I*:

> In April, 1950, the Committee distributed mimeographed copies of a proposed revision of the psychiatric nomenclature to approximately 10% of the membership of the American Psychiatric Association. Addressees were picked from the geographical listing of members, 10% of the members in each State and Canada being selected. In addition, addressees were selected by position held, in order to give complete coverage to all areas of psychiatry. Attention was paid to membership in other organizations (American Neurological Association, American Psychoanalytic Association, Academy of Neurology, American Psychopathological Association, etc.), so that a fair sampling of those groups was included. Members of the staffs of State Departments of Mental Health were included in order to obtain an expression of opinion from such departments concerning the statistical and clinical impact of the proposed revision.[7]

It appears that the emphasis was on bringing together psychiatrists with different specialties to include representatives of all areas in mental disorder research, as well as various interest groups of specialists in mental health. The foreword to *DSM-I* also states that the US Navy made some revisions to the work.[8] In short, the first edition of the *DSM* was the product of collaborations between the members of the APA, various medical professionals (psychiatrists and neurologists), and the military. These stakeholders, except for the military, remained the primary group of experts who revised the *DSM* and published *DSM-II*.

The psychoanalytic orientation of *DSM-I* and *DSM-II* was heavily criticized. Critics argued that it was based on theoretical presuppositions about the human mind (e.g., that the unresolved sexual tensions of childhood cause adult psychosis). Without observable and measurable evidence, they argued, the categories of mental disorders given in *DSM-II* failed the standards of scientific validity.[9] A scientifically valid category of mental disorder required external validators, such as symptoms (experienced by the patient), signs (observed by others), and neurobiological markers, not simply theories.[10]

Thus, two important changes occurred during the publication of *DSM-III*. First, the criteria for mental disorders were atheoretical, based on the signs and symptoms of mental disorders. Second, and more important for the purposes of this chapter, a procedural change made the revision process more collaborative. The DSM-III Task Force formed by the APA solicited input not only from psychiatrists but also from scientists in other stakeholder organizations, including the APA, the Association of Women Psychiatrists, the Academy of Psychiatry and the Law, the American Academy of

Child Psychiatry, the American Academy of Psychoanalysis, the American Association of Chairmen of Departments of Psychiatry, the American College Health Association, the American Orthopsychiatric Association, and the American Psychoanalytic Association. These groups received drafts of *DSM-III* and were invited to express their concerns.[11] Simply stated, these particular individuals and groups were selected because of their expertise. The introduction to *DSM-III* makes it clear that, whenever possible, the creators of the *DSM* sought the "advice of experts in each specific area under consideration."[12] Experts were understood to be trained researchers or clinicians who investigated mental disorders.

The creation process for *DSM-IV* remained largely identical to that for *DSM-III*; the DSM-IV Task Force and the various workgroups encouraged the participation of a wide range of professionals to serve as advisers. These advisers included individuals "from other health associations; clinical practitioners; researchers; forensic specialists; experts on gender, age, and cultural issues; and international experts."[13] From the publication of *DSM-I* to that of *DSM-IV*, patients were never consulted, although there were hints that this might change before the appearance of *DSM-5*.

Before the publication of *DSM-5*, there were extensive calls to the APA to include patients and other stakeholders in the revision process. These conversations took place largely in response to the DSM-5 Task Force's efforts to make the *DSM* creation process more collaborative. Some invoked the need for the process to be democratic, with participants including members of the public with a vested interest in the diagnostic criteria, such as patients and their families.[14] Some approached it from the perspective of patient advocacy and emphasized the need "for scientific experts to review their nosological recommendations in the light of rigorous consideration of consumer experience and feedback."[15] Others argued that patients bring a different perspective to the conversation on psychiatric classifications because they can report on their subjective experiences. This, they said, would help make the *DSM* criteria for mental disorders finer grained and more responsive to the real experiences of patients.[16] Still others argued that the *DSM* was facing a crisis of public trust and therefore needed to increase the integration of patients into the *DSM* revision process to address the issue.[17]

The DSM-5 Task Force members responded by acknowledging the benefits of including patients' perspectives in the *DSM* revision process, but they did not invite patients to join in. Their first set of worries was primarily epistemic: they raised concerns about the subjectivity of the data in patients' reports and the apparent conflict with psychiatry's desire to establish itself as an objective form of inquiry.[18] For example, in response to an article calling for the inclusion of patient-subjective data in the *DSM* and the *ICD*, they argued that "subjectivity is variable from person to person," making it "impossible to develop definitive criteria that would apply to every disorder." And they rhetorically asked, "Whose subjective experience should serve as a template by which criteria would be developed?"[19] The second set of worries was more practical in nature. The DSM-5 Task Force cited the limited time left in the revision process; there wasn't enough time to include patients. Instead, they suggested the APA's invitation for feedback from the public through an online forum was a positive step toward including patient input, insofar as patients are also members of the general public. Anyone, they said, was allowed to comment on *DSM-5*, including patients, and the task force would take this feedback seriously. To the best of my knowledge, however, no systematic report has detailed how such feedback was integrated into *DSM-5*—if at all. The following statement was released by the APA, in a document called "The People behind DSM-5":.

> Development of the fifth edition of *Diagnostic and Statistical Manual of Mental Disorders* (DSM-5) has been under way since 1999, when the American Psychiatric Association (APA) began recruiting a roster of diverse and internationally-recognized clinicians, scientific researchers and organizations to lend their expertise. Through this wide scale collaboration, the DSM-5 development process has involved not only psychiatrists, but also experts with backgrounds in psychology, social work, psychiatric nursing, pediatrics, and neurology. Participants in this effort represent 16 countries. All are leaders in their fields and are participating on a voluntary basis.[20]

Notably, the statement stresses the ongoing importance of recognized expertise.

In the end, the members of the APA (primarily psychiatrists) were the principal community of experts responsible for determining the criteria for the classification of mental disorders, and patients were once again left out of the process. Underlying this was the ongoing presupposition that patients did not have the expertise to participate in the scientific investigation of mental disorders, even though they might have something to offer as members of the public. Patients' possible contributions were discounted as "subjective" and thus incompatible with psychiatry's goal of being an objective science.

The explicit calls to include patients in the revision process of *DSM-5* were significant, but the participants in the debate took for granted that psychiatrists or other professionals trained in psychopathology were the experts and that including patient perspectives was important for reasons of added value (e.g., making the scientific process more democratic). Their involvement was not considered epistemically necessary in order to obtain complete knowledge on mental disorders. In other words, the arguments developed in favor of including patients were motivated by social concerns about making science more collaborative rather than epistemological concerns about enhancing the sources of information to develop a more comprehensive understanding of mental disorders. Even if the information patients provided on mental disorders was valued,[21] their perspective remained ancillary to that of trained experts.

Departing from those scholars who defend the necessity of patients' inclusion for social and political reasons, I call for the inclusion of patients in the scientific research on mental disorders for epistemological reasons. Their perspective is indispensable in the creation of knowledge on mental disorders, as I discuss in the third section of this chapter. One worry that must be addressed here is expressed by the DSM-5 Task Force when they discount patients' perspectives as too "subjective" in a science that aims to enhance its "objectivity." More specifically, the concern is that subjective data are interpersonally variable (not every person's testimony is like another person's), leading to the problem of whose perspective will count in developing the scientific framework for mental disorders. Critics argue that these two connected problems might end up compromising psychiatry's status as an objective science.

The notion of objectivity operative in these types of criticisms has been insufficiently examined, but tackling it is challenging, given the complexity of the notion.[22] The DSM-5 Task Force members seem to be construing objectivity as "impartiality," with objective data on mental disorders coming from third-person observations (by psychiatrists or other clinically trained experts). The construal of objectivity as "impartiality," possible only when data come from third-person perspectives, can be extrapolated from the assertions that "subjectivity is variable from person to person," making it "impossible to develop definitive criteria that would apply to every disorder."[23] Similarly, the rhetorical question asking "whose subjective experience should serve as a template by which criteria would be developed" hinges on the idea that criteria for mental disorders should be based on impartial or third-person observations, not "subjective" or first-person perspectives.[24]

The notion of objectivity as impartiality concerns me. It seems to rely on a now dated positivistic notion of objectivity as the "view from nowhere." Very briefly, according to logical positivists, science is objective to the extent that it is completely free from values. This is possible because, for positivists, justification for scientific evidence is grounded on impartial (third person) observational data that are verified by the world. Thus, objective science is impartial and abides by value neutrality. Feminist epistemologists have challenged the view of objectivity as the view from nowhere; rather, they argue, scientists are people, and science is a community activity.[25] However, even though the value neutrality thesis is false and science is guided by epistemic and contextual values, science can still be objective. The social feature of science is necessary for the objectivity of scientific inquiry. In fact, including a diversity of knowers in the knowledge production process is one way to enhance the objectivity of knowledge.[26] Science is objective to the extent that its methods, especially its criteria for assessing theories, are neither arbitrary nor subjective. In order to ensure objectivity in science, we must consider it as a social activity organized to permit and encourage *transformative criticism*—criticism with the power to change contextual values of scientists, should they be ill founded. As Longino writes:

> From a logical point of view, if scientific knowledge were to be understood as the simple sum of finished products of individual activity, then not only would there be no way to block or mitigate the influence of subjective preference but scientific knowledge itself would be a potpourri of merrily inconsistent theories. Only if the products of inquiry are understood to be formed by the kind of critical discussion that is possible among a plurality of individuals about a commonly accessible phenomenon, can we see how they count as knowledge rather than opinion.[27]

I would like to respond to the critics who argue that the inclusion of patients in the *DSM* revision process will compromise psychiatry's goals as an objective science by evoking Longino's response to positivists. Patients must be included in the knowledge production process for epistemic reasons; their inclusion will not compromise psychiatry's goal to be an objective science. The subjectivity of their experience with mental disorders and their perspectives on the diagnostic criteria are assets, not impediments, in the scientific process. Their direct encounter with mental disorders is an experience unknown to clinical or scientific experts. Including patients in the knowledge production process will allow transformative criticism by ensuring that different points of view are part of the scientific decision-making process. This, in turn, will lead to procedural objectivity, as the clinically trained experts and the patients can subject each other's hypotheses and evidential reasoning to critical scrutiny, thereby limiting the intrusion of individual subjective preferences into scientific knowledge. In the process, the subjective experiences of patients may create objective patterns of experiences previously missed by the clinically trained experts.[28] Of course, arguing for the indispensability of patients' perspectives in producing scientific knowledge requires a close assessment of why their points of view are so important; therefore, in the following section, I make a case for seeing patients as experience-based experts and for the epistemic value of their subjective reports.

Reevaluating Expertise in Psychiatry

As I explained in the previous section, the notion of expertise used in the psychiatric context is limited to those with training in psychopathology, but we need to reevaluate the epistemological question about who is qualified to conduct research and thus expand knowledge in psychiatry. I challenge the dominant notion of expertise in psychiatry. Patients, I argue, have a special kind of expertise when it comes to mental disorders, and they could contribute to psychiatric epistemology as "experience-based experts."[29] In addition, taking them as experience-based experts and including them in the knowledge production process in psychiatry overcomes the challenge of "subjectivity" put forward by the creators of the *DSM*. It is precisely these patients' subjective experience with mental disorders that makes them experience-based experts, and this perspective allows them to make a unique contribution to the understanding of mental disorders.

An important topic in philosophy of science and in science studies is the situation of scientific expertise in broad social, historical, and philosophical contexts.[30] A core question is whether the public, specifically those persons directly influenced by scientific research, should be involved in making scientific decisions or whether this should be left to the experts trained in the relevant sciences. A significant contribution to this debate has been the introduction of categories of expertise in science. The value of scientists' and technologists' knowledge and experience has been compared with others' knowledge and experience. In this new understanding, members of the public with special technical expertise by virtue of experience that is not recognized by degrees or other certifications are defined as "experience-based experts."[31] This enables *pockets of expertise* to be assigned to the citizenry. The goal is to develop a discourse of expertise that will place citizens' expertise alongside scientists' expertise. The framework of experience-based expertise can be used to characterize patients' encounters with mental disorders and their contributions to the scientific process. This raises the question of what expertise patients have.

Patients as Experience-Based Experts: Insights from the Natural Method

First-person reports of individuals with mental disorders offer unmatched resources for psychiatric epistemology, especially in the context of the current crisis and controversy in mental health research.[32] Before continuing, let me first list a few aspects of mental disorders on which patients have expertise. First, patients, especially those who have been dealing with a chronic mental disorder for a long time, have a robust understanding of the orientational challenges associated with mental illness, such as psychic pain or other symptoms and signs of the illness.[33] Second, they have knowledge of the various responses to different treatment methods, such as psychotropic medications, psychotherapy, and support groups.[34] Third, they understand how mental illness affects other aspects of their lives, such as interpersonal relationships, their treatment within the community, and the reactions of others to their illness.[35] Fourth, they understand how the scientific framing of their mental disorder (e.g., their *DSM* diagnosis) affects the kinds of accommodations they receive in their jobs and other aspects of their lives. Finally, those who have been successful in sustaining a good quality of life despite their illness have a good grasp of the sources of courage and strength they need to cope with disruptions in their daily lives.[36] Given their wealth of understanding, the inclusion of patients (i.e., those most directly influenced by scientific research programs and clinical treatments) can only benefit the process of knowledge production in medical epistemology.

The methodology proposed here, taking patients as experience-based experts to enhance knowledge on mental disorders, follows Owen Flanagan's natural method of solving what he calls the "puzzle of consciousness."[37] The natural method corrals consciousness by paying attention to its phenomenology (how consciousness seems), its psychology (what the mental labor of consciousness does), and its neurobiology (how consciousness is realized). In Flanagan's view, consciousness is a superordinate term for a heterogeneous array of mental state types. Adopting Thomas Nagel's description of the phenomenal character of consciousness from his seminal article "What Is It Like to Be a Bat?," Flanagan argues that all the mental types individuating consciousness share the property of "being experienced" or "being experiences," or, in other words, of there being "*something that it is like for the subject* to be in one of these states."[38] Flanagan says that, to build a theory of consciousness, we must deploy "the natural method" of coordinating all relevant informational resources at once, especially phenomenology, cognitive science, neuroscience, and evolutionary biology. By being responsive to both scientific findings and first-person experiences, this picture of consciousness is grounded on actual human selves as we encounter them.

In his later work on the problems associated with self-consciousness, dream interpretation, and narratives, Flanagan further develops the natural method, calling it "the expanded natural method."[39] In addition to looking at phenomenology, psychology, and neurobiology for insights into human consciousness, the expanded natural method draws insights from religion, literature, and other disciplines in the humanities. Such a multidisciplinary approach to human consciousness, Flanagan insists, demands constant vigilance against privileging or diminishing the value of any one approach.

My claim that patients are experience-based experts applies the natural method to psychiatric epistemology. In this trilateral strategy, scientific and clinical work involving individuals with mental disorders are the first two pillars of knowledge on mental disorders, and first-person reports are the third.[40] Taken together, these epistemic resources may help in developing effective interventions into mental disorders, as they disclose, respectively, the underlying causes of mental disorders, their effective treatment, and what it is like to have a mental disorder. The latter epistemic source can guide interventions in psychiatry, because firsthand accounts help in assessing therapeutic successes and failures. The information obtained is not exhaustive; rather, it is complementary to the information obtained from clinical and scientific research. Nonetheless, the inclusion of patient perspectives allows us to engage with different types of evidence that might be overlooked in conventional case-reporting scenarios.[41] Simply stated, their inclusion is necessary if we want to obtain a complete understanding of mental disorders.

The outcome of Flanagan's natural method to understand human selves is the embracing of human variations. Flanagan introduces the concept of "non-standard phenomenologies" to capture this diversity.[42] By that he means the wide range of capacities and limitations of human selves, some of which point to the distortion of the bonds connecting a person to her cognitive, bodily, social, and agential experience. His rich account of selfhood recognizes the existence of those with mental disorders and other cognitive diversities and includes them in the philosophical contemplation of how human selves should live their lives. Flanagan puts his naturalistic research program on human selves to the service of ethics and human flourishing. The recognition of the complexity of the psychological capacities and limitations of human beings by taking the scientific scrutiny of the actual lived world as simply one reference point results in naturalistic ethics.

Flanagan's recent work on addiction showcases the pertinent role played by the first-person perspectives of those with addiction in enhancing our understanding of this condition.[43] He writes that each of the scientific, clinical, and first-person perspectives contributes to the complete truth of alcoholism or alcoholisms.[44] Paying attention to all perspectives is necessary to understand any "psychobiosocial phenomenon, a social kind, not a natural kind, which has phenomenological, behavioral, social, cultural, genetic, and neurophysiological aspects or features."[45] Flanagan writes:

> The alcoholic who attempts to speak truthfully about what being in the grip is like, even if he does not get it exactly right even about himself, deserves a say in determining what alcoholism is because alcoholism is a complex psychobiosocial disorder with an irreducible first-person what-it-is-to-be like-ness. DSM-IV, and the soon-to-be-fully-unveiled DSM-5, speak truthfully, as far as we know, about the general contours of alcoholism and alcohol abuse given the data. But if one wants to know what it feels like to be an alcoholic, one is much better off reading, say, Charles Jackson's Lost Weekend (1944), the best description ever of the phenomenology of alcoholic craving, scheming, the multiple kinds of alcoholic forgetfulness, forgetfulness of what one said and did, of the alcoholic's self-degradation, self-loathing, of the abject awfulness of his predicament, the suicidal despair.[46]

Drawing on Flanagan's natural method, we can imagine a trilateral strategy to investigate mental disorders by including patients among the experts on them.[47] This will have scientific significance because of its ability to capture mental disorders in their full complexity. At the same time, it will have ethical implications, affecting how we treat those with mental disorders and contributing to their flourishing. The inclusion of patients will result in scientifically grounded ethical and policy-related initiatives that can improve the lives of those suffering from mental disorders.

Looking at patient reports by using the natural method may help us understand mental illness and facilitate *DSM* development, but does it mean patients themselves ought to be involved in designing the *DSM*? This question is at the heart of this chapter. As I see it, any reporting on the first-person perspective by others will necessarily be selective and representative and will not capture the full fabric of what the patients themselves are reporting. For example, when clinicians report on what their patients' experiences are like, they are selective; they are necessarily influenced by their own perceptions of the patients and their theoretical orientations. What we get directly from the first-person reports, in the *DSM* designation process, is raw input reflecting patients' standpoints. This would be missed if we relied on others' reports of first-person perspectives.

A second question is whether my view of including patients in the knowledge production process in psychiatry applies to other areas of medicine. I say it does. We already see examples of this line of thinking in all areas of medicine; there are many, but let me give two here. First, in the last few decades, an important topic for proponents of humanistic approaches to medicine has been the nature of the relationship between the health care professional and the patient. One suggestion to enhance humanism in medical practice is for health care professionals to recognize and engage with patients' narratives. Rita Charon, the founder of the narrative medicine approach, argues that the clinician must acquire the skills to listen to, interpret, and reflect on the patient's stories with an "engaged concern" to achieve therapeutic outcomes, because this is the fundamental way in which the patient learns to trust the clinician.[48] The focus of narrative medicine is on the experiences of the patients rather than on the generalizable propositions about them produced by logicoscientific inquiry. Giving uptake to patients' narratives is considered necessary not only to build trust between clinicians and patients but also to give physicians the means to improve the effectiveness of their work with patients, themselves, their colleagues, and the public. Active research and clinical programs in various hospitals is testing the fundamental tenets of narrative medicine in medical practice. My proposal to include patients in the knowledge production process in psychiatry is developed in a similar spirit of humanism: one of the best ways of enhancing psychiatric epistemology and practice is to give a voice to those directly affected by mental illness.

The second example of recent advocation for the direct inclusion of patients' perspectives in the medical knowledge production process is the current discussion of the nature of clinical case studies in medical epistemology. A clinical case study, as a research methodology, is an empirical inquiry that investigates a medical phenomenon within its real-life context. Clinical case studies are based on in-depth investigations of individuals' illnesses to describe the details of a case, explore its underlying causes, study its unique aspects, create effective interventions, and so on. Rachel Ankeny recently noted that most clinical case studies are only presented from the physician's point of view, in which "the doctor's voice becomes authoritative, even though in a sense his or her version of the events could be seen as a mere interpretation of the 'real' case as narrated by the patient."[49] She argues for the inclusion of patient perspectives in clinical studies not just as a source of added value but also as a source of evidence, because patients provide unique details about their illness that may otherwise be overlooked.[50]

Conclusion

In this chapter I have argued that patients have experience-based expertise on mental disorders and thus must be included in the scientific research programs advancing knowledge about these disorders. Patients' involvement as contributors to research has been minimal because of the presupposition that they lack the technical expertise required to make meaningful contributions to psychiatric science. Acknowledgment of their possible contribution has been limited to their position as members of the general public with a certain vested interest in the research. All they can offer is subjective information, however, and this is incompatible with psychiatry's goal of establishing itself as an objective science. Partially inspired by Flanagan's natural method, I argue that the very subjectivity of the experience of patients makes them experience-based experts on mental disorders and provides indispensable resources for establishing psychiatry as an objective science. The first-person reports of those suffering from mental disorders represent a pillar of psychiatric epistemology, right beside those of scientific and clinical research, and should be acknowledged as such. The stakes are high, not only in scientific and medical terms but also in ethical terms.

Notes

1. American Psychiatric Association, *Diagnostic and Statistical Manual of Mental Disorders, Fourth Edition*, xxiii; American Psychiatric Association, *Diagnostic and Statistical Manual of Mental Disorders, Fifth Edition*, xxvi.

2. American Psychiatric Association, *Diagnostic and Statistical Manual of Mental Disorders, First Edition.*

3. Ibid.; Greenberg, Shuman, and Meyer, Unmasking forensic diagnosis.

4. American Psychiatric Association, *Diagnostic and Statistical Manual of Mental Disorders, First Edition.*

5. Ibid., 9.

6. Ibid.

7. Ibid., viii–ix.

8. Ibid.

9. Beck, Reliability of psychiatric diagnoses; Schwartz and Wiggins, Diagnosis and ideal type; Tsou, DSM-5 and psychiatry's second revolution; Sadler, *Values and Psychiatric Diagnosis.*

10. Robins and Guze, Establishment of diagnostic validity in psychiatric illness; First et al., Clinical utility as a criterion for revising psychiatric diagnoses.

11. American Psychiatric Association, *Diagnostic and Statistical Manual of Mental Disorders, Third Edition.*

12. Ibid., 4.

13. American Psychiatric Association, *Diagnostic and Statistical Manual of Mental Disorders, Fourth Edition*, 851.

14. Sadler and Fulford, Should patients and their families contribute to the DSM-5 process?

15. Stein and Phillips, Patient advocacy and DSM-5.

16. Flanagan, Davidson, and Strauss, The need for patient-subjective data in the DSM and the ICD.

17. Bueter, Public epistemic trustworthiness and the integration of patients in psychiatric classification.

18. Regier et al., Patient involvement in the development of DSM-V.

19. Ibid.

20. American Psychiatric Association, *Diagnostic and Statistical Manual of Mental Disorders, Fifth Edition.*

21. For example, see Flanagan, Davidson, and Strauss, The need for patient-subjective data in the DSM and the ICD.

22. Longino, *Science as Social Knowledge*; Douglas, The irreducible complexity of objectivity.

23. Regier et al., Patient involvement in the development of DSM-V.

24. Ibid.

25. Longino, *Science as Social Knowledge*; Douglas, The irreducible complexity of objectivity.

26. Longino, *Science as Social Knowledge.*

27. Ibid., 74.

28. Ibid. Longino, in her discussion of transformative criticism, argues that scientists should be the ultimate decision makers, albeit a diverse community of scientists, representing different viewpoints and schools of thought. Longino herself therefore may not advocate including patients in the scientific decision-making process, as they are not scientists. In defending patients' inclusion in the knowledge production process from the standpoint of feminist epistemology, I invite readers to think of transformative criticism as a process that includes not only technically trained scientists but also other kinds of experts, such as patients, who are experience-based experts, as I discuss in the third section of this chapter. The kind of objectivity I envision could thus be labeled participatory intersubjective objectivity. It is beyond the scope of this chapter to develop this view.

29. Collins and Evans, The third wave of science studies.

30. Ibid.

31. Ibid., 251.

32. Poland and Tekin, *Extraordinary Science and Psychiatry*; Tekin, Flanagan, and Graham, Against the drug cure model; Tekin and Outram, Overcoming mental disorder stigma.

33. Tekin, Self-insight in the time of mood disorders.

34. Tekin, The missing self in scientific psychiatry.

35. Tekin, Self-concept through the diagnostic looking glass; Tekin, Self-insight in the time of mood disorders.

36. Tekin and Outram, Overcoming mental disorder stigma.

37. Flanagan, *Consciousness Reconsidered.*

38. Ibid.; Nagel, What is it like to be a bat?

39. Fireman, McVay, and Flanagan, Introduction.

40. Tekin, Are mental disorders natural kinds?

41. Ankeny, The role of patient perspectives in clinical case reporting.

42. Flanagan, *Consciousness Reconsidered.*

43. Flanagan, Identity and addiction, 865–888; Tekin, Flanagan, and Graham, Against the drug cure model.

44. Flanagan, Identity and addiction.

45. Ibid., 866.

46. Ibid., 867.

47. Tekin, Are mental disorders natural kinds?

48. Charon, *Narrative Medicine.*

49. Ankeny, The role of patient perspectives in clinical case reporting, 102.

50. Ibid.

Bibliography

American Psychiatric Association. *Diagnostic and Statistical Manual of Mental Disorders, First Edition: DSM-I.* Washington, DC: American Psychiatric Association, 1952.

American Psychiatric Association. *Diagnostic and Statistical Manual of Mental Disorders, Second Edition: DSM-II.* Washington, DC: American Psychiatric Association, 1968.

American Psychiatric Association. *Diagnostic and Statistical Manual of Mental Disorders, Third Edition: DSM-III.* Washington, DC: American Psychiatric Association, 1980.

American Psychiatric Association. *Diagnostic and Statistical Manual of Mental Disorders, Fourth Edition: DSM-IV.* Washington, DC: American Psychiatric Association, 1994.

American Psychiatric Association. *Diagnostic and Statistical Manual of Mental Disorders, Fifth Edition: DSM-5.* Washington, DC: American Psychiatric Association, 2013.

Ankeny, R. A. The role of patient perspectives in clinical case reporting. In *Knowing and Acting in Medicine*, edited by R. Bluhm, 97–111. Lanham, MD: Rowman and Littlefield, 2017.

Beck, A. T. Reliability of psychiatric diagnoses: A critique of systematic studies. *American Journal of Psychiatry* 119, no. 3 (September 1962): 210–216.

Bueter, Anke. Public epistemic trustworthiness and the integration of patients in psychiatric classification. *Synthese* (September 2018): 1–19. https://doi.org/10.1007%2Fs11229-018-01913-z.

Charon, Rita. *Narrative Medicine: Honoring the Stories of Illness*. New York: Oxford University Press, 2006.

Collins, H. M., and Evans, R. The third wave of science studies: Studies of expertise and experience. *Social Studies of Science* 32, no. 2 (2002): 235–296.

Douglas, Heather. The irreducible complexity of objectivity. *Synthese* 138, no. 3 (February 2004): 453–473.

Fireman, Gary, McVay, Ted, and Flanagan, Owen. *Narrative and Consciousness: Literature, Psychology and the Brain*. Oxford: Oxford University Press, 2003.

First, M. B., Pincus, H. A., Levine, J. B., Williams, J. B. W., Ustun, B., and Peele, R. Clinical utility as a criterion for revising psychiatric diagnoses. *American Journal of Psychiatry* 161, no. 6 (June 2004): 946–954.

Flanagan, E. H., Davidson, L., and Strauss, J. S. The need for patient-subjective data in the DSM and the ICD. *Psychiatry* 73, no. 4 (2010): 297–307. https://doi.org/10.1521/psyc.2010.73.4.297.

Flanagan, Owen. *Consciousness Reconsidered*. Cambridge, MA: MIT Press, 1992.

Flanagan, Owen. Identity and addiction: What alcoholic memoirs teach. In *The Oxford Handbook of Philosophy and Psychiatry*, edited by K. W. M. Fulford, M. Davies, R. Gipps, G. Graham, J. Sadler, G. Stanghellini, and T. Thornton, 865–888. Oxford: Oxford University Press, 2013.

Greenberg, S. A., Shuman, D. W., and Meyer, R. G. Unmasking forensic diagnosis. *International Journal of Law and Psychiatry* 27, no. 1 (February 2004): 1–15.

Longino, Helen. *Science as Social Knowledge: Values and Objectivity in Scientific Inquiry*. Princeton, NJ: Princeton University Press, 1990.

Nagel, Thomas. What is it like to be a bat? *Philosophical Review* 83, no. 4 (October 1974): 435–450.

Poland, Jeffrey, and Tekin, Şerife, eds. *Extraordinary Science and Psychiatry: Responses to the Crisis in Mental Health Research*. Cambridge, MA: MIT Press, 2017.

Regier, Darrel, Kuhl, Emily, Kupfer, David J., and McNultry, James. Patient involvement in the development of DSM-V. *Psychiatry: Interpersonal and Biological Processes* 73, no. 4 (2010): 308–310.

Robins, E., and Guze, S. B. Establishment of diagnostic validity in psychiatric illness: Its application to schizophrenia. *American Journal of Psychiatry* 126, no. 7 (1970): 983–987.

Sadler, J. Z. *Values and Psychiatric Diagnosis*. Oxford: Oxford University Press, 2005.

Sadler, J. Z., and Fulford, B. Should patients and their families contribute to the DSM-5 process? *Psychiatric Services* 55, no. 2 (February 2004): 133–138.

Schwartz, M. A., and Wiggins, O. P. Diagnosis and ideal type: A contribution to psychiatric classification. *Comprehensive Psychiatry* 28, no. 4 (1987): 227–291.

Schwartz, M. A., and Wiggins, O. P. Typifications: The first step for diagnosis in psychiatry. *Journal of Nervous and Mental Disease* 175, no. 2 (February 1987): 65–77.

Stein, D. J., and Phillips, K. A. Patient advocacy and DSM-5. *BMC Medicine* 11, no. 133 (May 2013). doi: 10.1186/1741-7015-11-133.

Tekin, Şerife. Are mental disorders natural kinds? A plea for a new approach to intervention in psychiatry. *Philosophy, Psychology, and Psychiatry* 23, no. 2 (2016): 147–163.

Tekin, Şerife. The missing self in scientific psychiatry. *Synthese* 196, no. 6 (2019): 2197–2215.

Tekin, Şerife. Self-concept through the diagnostic looking glass: Narratives and mental disorder. *Philosophical Psychology* 24, no. 3 (March 2011): 357–380.

Tekin, Şerife. Self-insight in the time of mood disorders: After the diagnosis, beyond the treatment. *Philosophy, Psychiatry, and Psychology* 21, no. 2 (2014): 139–155.

Tekin, Şerife, Flanagan, Owen, and Graham, George. Against the drug cure model: Addiction, identity, and pharmaceuticals. In *Philosophical Issues in Pharmaceutics: Development, Dispensing, and Use*, edited by Dien Ho, 221–236. New York: Springer, 2017.

Tekin, Şerife, and Outram, Simon. Overcoming mental disorder stigma: A short analysis of patient memoirs. *Journal of Evaluation in Clinical Practice* 24, no. 5 (July 2018): 1114–1119.

Tsou, Jonathan. DSM-5 and psychiatry's second revolution: Descriptive vs. theoretical approaches to psychiatric classification. In *The DSM-5 in Perspective: Philosophical Reflections on the Psychiatric Babel*, edited by S. Demazeux and P. Singy, 43–62. Dordrecht: Springer, 2015.

4 Addiction and Ultimate Concern: Flanagan Paired

George Graham

Introduction

Often a philosopher's defense of a particular position on one topic or problem is inseparable from his or her position on another topic, and so on for several other topics.[1] Philosophical inquiries into the nature and significance of addiction vividly exhibit this pattern of interconnected analyses. The problem of understanding addiction is not one topic or problem but several: the problem of rationality, the problem of self-control, the issue of intertheoretic explanatory reduction to neuroscience, the question of the nature of unwilling addiction and of relapse, and so on.[2]

The present chapter outlines in broad strokes some features of Owen Flanagan's work on addiction. I plan to connect his work on this topic with other topics he has discussed and to complement his work by making reference to certain other ideas—to pair his work with that of other workers. Ideas of my own enter into the mix as well.

To honor Owen Flanagan is to honor his vision of the aims of philosophy and the character of philosophic progress. In the case of this chapter, the honoring consists of an illustration of his conception of philosophical theory construction and of the importance of intellectual social cooperation, which, for him, is essential to a construction's proper exercise.

Unwilling Addiction

On the topic of addiction, Flanagan has been mostly, but not exclusively, interested in alcoholism as a type of addiction, which, as he puts things, can be "woven into the fabric of an alcoholic's life."[3] "There are many ways," he writes, in which consuming alcohol "can gain entry into the fabric of one's being."[4]

Many alcoholics want to cease being alcoholics. They want alcoholism out of their life. Flanagan's focus on addiction/alcoholism is aimed at addicts or alcoholics who want to cease being addicted. These are so-called *unwilling* addicts. An unwilling addict fights or struggles against their addictive behavior patterns and the negative roles of those patterns in their personal life.[5]

Getting addictive alcohol consumption out of a person's life, Flanagan cautions, requires an unwilling alcoholic to recognize that he or she "bears a mother lode of responsibility for [doing so]."[6] This is no small chore, for an unwilling addict is locked into what Flanagan calls a "performative inconsistency," leading a life that he or she knows runs harmfully contrary to their own health and best interest.[7] An unwilling addict is "interested in ... self-control, but [has] trouble leveraging that control."[8]

The life-woven messes of unwilling addictions puzzle Flanagan. It is one puzzle that interests both of us.[9] The puzzle is categorical or taxonomic, and I phrase it in the form of a question: Are unwilling addictions defects of the brain—neurological disorders of a sort? Parkinson's and Alzheimer's are defects or disorders of the brain. So, the questions are: Are unwilling addictions—alcoholism, for example—in the same general taxonomic category or box? And are they best classified, understood, and treated as brain defects?

In proposing an answer to these questions, which I will do in a moment, I make two background assumptions. I do not defend these assumptions here, but each of them, I presume, is more or less uncontroversial.

The first assumption is that a brain defect is a type of flaw or dysfunction that is not a product of the fact that we human beings deem or believe it so but is a way in which the brain is not supposed to be. This is a way that is, in an important sense, objective in a manner that is analogous to a defect of the heart if it cannot pump blood or of limbs that cannot move.

The second assumption is that a brain defect, properly so called, is not a defect or flaw in evolution's or Mother Nature's design of the brain and central nervous system. Our brains have naturally evolved or been selected to enable human beings to take risks as well as to learn on variable ratio schedules of reinforcement. Risk taking makes recreational drug consumption possible. Variable ratio schedules shift into high gear at gambling casinos and racetracks. These activities may be harmful to a person and, in some cases, are not under proper or prudential reason-responsive self-control and qualify as addictions, although in and of themselves they are not expressions of defects in neural design. It is not as if Mother Nature set out to build or select a brain that is immune to addiction. Alcoholism/addiction is not a failure of design.[10]

Addiction and the Brain Defect Thesis

Different terms or concepts are used to identify brain defects or disorders of the types deemed responsible for an addiction. Defect types are sometimes described as neural systems or subpersonal connections between systems that are "diseased," "damaged," or "broken."[11] Although the terminology may differ, the main idea, in general, behind a case of brain defect is that the causal powers or dispositions of a defective (diseased, etc.) brain are significantly different from those of a nondefective brain. So, a complete picture of a brain that is defective and therein explanatorily responsible for an addiction must describe the nature of the defect and its explanatory role or roles in addictive behavior.

Concerning the specific identity of the relevant brain defect, some theorists claim that addiction is a defect of the brain's mesolimbic dopamine reward system.[12] The condition is, as some advocates put it, a case of a "highjacked brain."[13] The term *hijack* is a metaphor, of course. So understood, if an addiction is a type of brain defect, then there is something pathologically wrong with, say, the operation of the midbrain reward or dopaminergic system.[14] Just as there are, say, neurological motor disorders in the ability to initiate, execute, and terminate bodily movement (in, say, Parkinson's disease), there also are neural defects or disorders in reward processing. Addictions are conditions with such defects, or so some investigators claim.

I am critical of classifying addiction as a brain defect,[15] but not because I don't believe any types of addiction may qualify as brain defects. Nor am I critical because I deny that addictions have neurobiological bases or substrata. Some cases of addiction may represent brain defects. And, of course, no case of addiction is without substrata.

No, I am critical because, I believe, and as I argue elsewhere, that a person's inhibitory self-control may be severely compromised, impaired, or derailed in manners that are distinctive of (unwilling) addictive behavior but without being defective or flawed neurologically.[16] The reward system, for example, may not be diseased or broken. It may just be helping to undergird an unhealthy form of prudentially uncontrollable behavior.[17]

An analogy may help to explain what I mean when I refer to a defect-free or unbroken neural base or substratum for an addiction. Imagine the following situation. My front doorbell broke last month. The electric circuit inside the bell was damaged, so I purchased and installed a new bell. The new one was designed just like the old one. It worked well for a week. However, a neighborhood squirrel has become enamored of the bell. Using a railing on the front stoop, he climbs up to the bell's button and, for whatever reason, pushes his paw against it. The bell rings inside the house. In fact, the squirrel (much like the postman[18]) always rings twice. This is annoying to me and to my family.

The squirrel-based difficulty with the bell is not a mechanical defect per se, as it was with the busted circuit. So, installing yet another new bell of the same type or design will not solve the problem or discourage the squirrel.

I am tempted to dub the bell's unwelcome pattern of ringing, in honor of Ulric Neisser and J. J. Gibson, as an ecologically situated doorbell disorder.[19] By this I mean that it is a disorder in the manner in which the bell is situated in, and behaves in response to, specific types of events in its physical environment. Ecologically speaking, the bell is situated at my home's portal, and this somehow encourages the squirrel.

Talk of an ecologically situated disorder would be meant to identify a disorder or unwelcome behavior in which references to environmental factors are part of the explanatory story of the onset and persistence of the behavior—rather than just references to processes within a relatively isolable physical structure (such as a bell). Note that in order to account for the "situated" bell's ringing in an unwelcome way or manner, I need to refer to its electric circuit and to the spring in the bell's button. But, of course, there is something else to which I must refer: the behavior of the squirrel. The squirrel is a factor external to the bell as a mechanism.

Questions concerning environmental factors in addictive behavior patterns are particularly germane to understanding the failure of an (unwilling) addict to properly or consistently leverage his or her desired forms of self-control, not just context or circumstance as a background to addiction but rather as determining factors of an illness or disorder. A full explanatory story of an addiction may require reference to stressful social events, emotional biases connected with social circumstances, peer groups, culturally biased forms of attention and memory, the social consequences of failure of general attempts to refrain from the addictive activity, and a host of other environmental contingencies.[20]

To be sure, situational or environmental inputs into bells and brains are not the same general kinds of things. Bells respond to pressures on their buttons. Human brains respond to immensely more subtle and complex features of their social and cultural circumstances.[21] Likewise, the squirrel case is a case of being competent at bell pushing, but without the animal's comprehension or understanding of bell pushing. Squirrels don't know that they are pushing buttons on doorbells. Human beings, by contrast, have a huge and spectacular range of comprehended competencies. Libraries are filled with displays of our erudite understanding, not just of doorbells, of course, but of people and the world in general. Unwilling addicts know that their behavior is harmful, and they wish to put a stop to it.

On a number of occasions in which I have argued that addiction is not necessarily the result of a brain defect, I have encountered the following counterargument: "Addiction's path courses through or is underwritten by activities in the brain and central nervous system. This fact alone is enough to warrant the proposition that an addiction is a type of brain defect."

The objection misses the point of the nondefect claim, as may be shown by mounting an analogous point about the phenomenon of morning sickness. In cases of bodily sickness, we do not automatically assume that something is wrong with a person's bodily health and therefore there is a somatic defect just because he or she feels ill or is sick. Morning sickness, for example, transpires in or courses through the body of a pregnant woman, but this does not necessarily mean that there is something biologically or mechanically wrong with her body. Morning sickness may be an evolved biological defense mechanism against the ingestion of fetal toxins.[22] If or when its symptoms assume life-threatening forms or degrees (such as with heavy vomiting and severe dehydration), morning sickness requires medical treatment. However, this fact alone does not imply that morning sickness per se is a somatic defect or bodily dysfunction. Fever, too, however unpleasant and at times in need of medical address, sometimes is a defense mechanism against infection,[23] not a breakdown of bodily health. So, to reason by analogy, the mere fact that the pathway of addiction courses *in* or through the brain does not entail that addiction is a disorder *of* the brain, meaning a brain or neurological defect. More than an eagerness to acknowledge the neural pathways of addiction is needed to argue that scientific hypothesis or claim.

G. Lynn Stephens and I once put matters with respect to drug addiction and the causal role of the midbrain as follows:

> Should we conclude ... that drug addiction represents a disease or disorder in the brain reward system? That would be a very hasty inference. The same brain reward system sub-serves a whole variety of psychological and biological functions: notably learning and response to injury or other sorts of stress. In decreasing the dopamine reward for repeated instances of the same stimulus, the brain may be functioning just as Mother Nature (natural selection) designed it to function. That there are specific neurological mechanisms that underlie the addictiveness of drug consumption and contribute to substance abuse is consistent with those mechanisms functioning adequately, neurologically speaking, insofar as they underlie the adaptive activities of learning and response to injury.... Facilitating prudent consumption of a recreational drug is simply not what the [brain reward] system was designed to do. So, from a neurobiological point of view, the brain might be functioning well, even if in doing so, the brain reward system underwrites an addict's mental disorder.[24]

The shift from describing addiction as a brain defect to appreciating that addiction may at least sometimes be empirically compatible with a neurologically healthy or normal brain helps to draw attention to the question of how an addictive behavior pattern may be woven into a person's life situation and require for its elimination a fairly labor-intensive reconstruction of who a person is. Numerous voluntary actions are required in order to become an addict. Many voluntary actions also are required in order to refrain from or abstain from being an addict, especially when countervailing environmental contingencies and psychological stresses or strains make abstention difficult.[25]

The fact is that describing relevant levels or types of explanations for any mental illness or disorder, addiction included, requires making delicate explanatory discriminations and discerning subtle relationships among potentially numerous factors both distal (such as an individual's learning history) and proximate (such as the stresses and strains of the person's current context). Addictions and other mental illnesses do not merely course through a brain. They course through the brain of a reason-responsive, socially or ecologically situated person or self-conscious agent.[26]

Ultimate Concern

While writing a draft of this chapter, my mind was not just on Owen Flanagan and his work on addiction. I was also thinking of a concept by Paul Tillich, a distinguished theologian whose ideas I had studied at my collegiate alma mater, Fordham University in the Bronx, the same alma mater as Owen's.

Tillich (1886–1965) wrote about something he called an "ultimate concern," by which he meant a deliberate, deep, self-conscious personal commitment that centers an individual's personality and holds his or her life together, so that it does not fracture or come apart at its volitional seams.[27] Tillich wrote: "Ultimate concern gives depth, direction and unity to all other concerns, and with them, to the whole personality."[28]

A person without an ultimate concern may possess several concerns, interests, or commitments that are distinct and are not ranked so that one, namely an "ultimate" concern, has supremacy and integrates or unifies the others. Any single concern may be abandoned, perhaps even without good reason. By comparison, a person possessing an ultimate concern maintains this particular concern, as Robert Neville describes Tillich's idea, "in the face of opposition when all other concerns have been given up."[29] Neville adds, "When the pressures of life demand choice between concerns, the one that is ultimate either rejects the others or subordinates and adapts them to itself."[30]

For Tillich, an ultimate concern formally or abstractly conceived is not restricted to religion or theology. All sorts of commitments, to family and friends, to making money, to politics, to a life of moral principle, to the arts, and so on, may be an ultimate aim or focus of a life.

We humans lead complex and complicated lives. When an individual becomes addicted, unwillingly so, and wants the condition out of their life, he or she may appreciate that, in the words of clinical psychologist Richard Bentall, "recovery [must involve] the development of new meaning and purpose in one's life."[31] A new ultimate concern is needed, something not just to displace but rather to replace addiction. This should be a commitment that precludes a willingness to engage in addictive consumption and reflects (to deploy Harry Frankfurt's terms) "necessities of the will [that] concern what [a person now is] unable to bring [himself or herself] to do."[32]

Death and Agentive Possibility

Addiction, of course, may be dangerous to a person. It may end an addict's life.

Death is the irrevocable end point to personal existence and agentive possibility. If you don't do something before you die, you are never going to do it.

Death reveals how limited life is. As Robert Nozick (1938–2002) put it, death sets the "absolute *finis* to our future," the finite sum to what we have done.[33]

Owen Flanagan has written about death. He believes that it places a person under a moral imperative to use his or her life to improve the human lot. He writes: "Here is how I think of my own death.... When you die *you* lose nothing you care about. Why? Because *you* are no longer there. That which is gone has nothing to lose.... But your good effects continue on. This is something to be proud of and happy about while you are alive."[34] Good consequences of a life may endure as the effects of a person's moral engagement, kindness, compassion, and generosity.

Imagine an unwilling and clinically depressed addict who believes that his or her life has been an utter waste or failure. A depressed addict may severely underestimate the sometime successes or good consequences of their life.[35] Suppose this particular addict has had and continues to have various morally good effects of their life. These are achievements to be proud of, such as compassion for other addicts, progressive forms of political engagement, friendships, or whatever.

Life is hard. Just because a person is addicted does not mean that his or her life is devoid of goodness. Although an alcoholic's achievements do not make them paradigms of personal and moral responsibility, for they are addicted and harm themselves and perhaps other people, they are not necessarily utter losers or failures as persons.

The assumption that an addiction demerits or makes a failure of a whole life neglects the limits for goodness inherent in all of us. A deeply impaired or marred life may still leave a significant measure of goodness in the world or, to put it somewhat differently, we must recognize that the goodness of a life comes in degrees. Talk of degrees may seem obvious, but that certainly does not make it empty or unhelpful. A valuable degree of goodness may attach to the life of an addict.

Flanagan never explicitly says so, at least as far as I know, but I suspect that he believes that certain formal features of addiction should be part of a postaddicted life. Some aspects of an addictive pattern may be wisely shared by postaddicted behavior, although excised from addiction's negative consequences and the often attendant feelings of shame and personal failure. One of those formal features is that just as addiction may structure or center a life, so may a recovered addict need something else to structure or center their life. That is, just as an addictive pattern may both cognitively capture and motivationally magnetize a person, a recovering or recovered addict may need something else to capture and hold them. This is an ultimate concern, as it were. It is something that marks a subject's personal space of meaning and significance and that sets parameters for personally acceptable behavior.

Flanagan describes the ultimate concern or life focus that he recommends, not just in response to an addiction but period. He praises a "sense of connection with that which is greater than and more comprehensive than [one's own] self."[36] This something for him is moral engagement or commitment. Moral engagement or commitment is a form of union of self with nonself or other selves. It abandons the struggle for private happiness and is bound compassionately to other people and to their suffering and welfare.[37]

Moral engagement accepts the facts of human finitude and temporal limits. An ethically committed person must often struggle courageously against those limits in order to keep from being too narrow in kindness, too tribal in sympathies, and too parochial in loyalties.[38]

The temporal poignancy of human finitude increases with age. If circumstances and time permit, a person reaches an age group in which "none of us can see as well, hear as clearly ... sleep as soundly as [their] younger and stronger selves."[39]

As we age, the value we place on our lives may seem to depend, in part, not just on what we once were able to do physically but can no longer do, but also on the lives or courses of life that we could have chosen but did not—on the opportunity costs of the paths, people, things, or alternatives we rejected, excluded, or neglected. However, in contrast to a perhaps melancholic emphasis on opportunities lost, an ethical perspective or vision asks us to picture ourselves in an expanding circle of *present* concern and *current* opportunity, not in terms of diminished competence or poignant regret.

Friend as Owen Flanagan is of an empirically enriched brand of Buddhism, he points out that in taking the bodhisattva's vows, a person should concentrate not on things that he or she could have done but failed to do, but rather on things that the person still can and should do. What are the opportunities or potentialities that I face at *this* moment? I cannot rewrite the past. That would be a myth or fairy tale. I can give myself over, however, to present opportunities for which the past is prelude, devoting myself to the present's moral possibilities for the future and to embracing current occasions for contributions to other people as an ultimate concern.[40]

Flanagan claims that we ought to care for the common good of the wider world and do so with an attitude of "greatly amplified fellow feeling. We ought to believe that everyone deserves to be free of suffering and to achieve some sort of happiness."[41]

Without God

What about God? The assumption that breaking out of the bounds of a selfish ego, and being inspired by moral engagement, requires a commitment to theism is both widespread and influential. As one writer expresses this popular conviction, "It is ... impossible for people to be moral without religion or God."[42]

The writer could not be more mistaken in the claim of impossibility. The cognitive development literature on childhood and moral attitude formation reveals that moral attitudes are present in people's minds quite independent of religion or of cultural or parental exposure to supernatural beliefs about a moralized god. As anthropologist Pascal Boyer remarks, "Moral intuitions [in children] appear long before children represent the powers of supernatural agents, [and] they appear in the same way in cultures where no one is much interested [in God or] the powers of supernatural agents."[43]

I do not wish to deny that a commitment to a conception of divine benevolence or to beliefs about divine rewards (and punishments) may be, as Paul Bloom puts it, accelerants to moral behavior;[44] not deciders or determinants of moral truths, but encouragements or reinforcements for morally proper actions. However, we should reject the notion that a godless morality is a contradiction in terms. People certainly may be moral without being theists.

Flanagan admits to no god. He conceives of the power of moral engagement, however, as something spiritual, not in a theological, God-tethered, or supernatural sense but in a reverential and aspirational sense. Responding to moral or ethical demands, he says, is hard work; and, when properly conceived, it is also compassionate, merciful, and kind—a demanding ideal. Ideals are like that. They make demands on us.[45]

For Flanagan, God is not, as such, a well-chosen part of the nonegoistic ethical picture. Why not? In short, this is because, as Flanagan says, "Modern science and theism ... don't ... sit together comfortably."[46] For him, theistic belief does not square with the best scientific understanding that we have of the world. Indeed, the concept of a god or supernatural agent, who created the world, is an explanatory nonstarter as far as he is concerned. For Flanagan, there are no facts about the world that are better explained in theistic rather than in nontheistic and naturalistic terms, when explainable at all.[47]

Natural Method and Ways of the World

Rejection of theism on grounds of empirical or explanatory vacuity raises a question about the nature and scope of empirical knowledge. This is the big epistemic question of how best to conceive of the natural world, devoid of divine influence. Is the natural world describable in one and only one way? Is there one and only one single, sound, and sensible ideal depiction of the world?

I assume, and so does Flanagan, that the answer to the question is negative. The natural world consists of the many ways, manners, or modes in which it is or can be truly seen, described, pictured, or spoken about. There is no single way in which the world is. Empirically, the world is not all of one type or univocal kind.

Flanagan is a descriptive and explanatory pluralist about the natural world and its ways and truths. Like any good pluralist, he is not unselective or arbitrary in his pluralism.[48] With all the academic fields, perspectives, and disciplines (the arts, the sciences) at work in describing the world, there is a *dilemma* about how to regiment, constrain, or order them and about which ways (descriptions, depictions, versions, explanations) to believe are true. The topic of whether, when, and how to select this or that version or description of things is a complex and often difficult empirical question.

While a readiness to recognize numerous ways of describing the world may be, as Nelson Goodman puts it, "liberating and suggestive of new avenues of exploration, a willingness to welcome all ways ... is no substitute for hard work."[49] He adds, "Not even a fly is likely to take one of its wings as a fixed point."[50] Pluralism must be directed or regimented. It must find the right pieces in "a jigsaw puzzle" that produce a proper fit and "increase acuity of insight [and] a range of comprehension."[51]

Flanagan's so-described *natural method* and application of this method across a range of topics from addiction, consciousness, and moral psychology to ethics and dreams constitutes a welcome contribution to a pluralistic understanding of the ways of the world. One of the method's main attractions is that "it can be expanded indefinitely as needed to bring to bear all the data necessary to explain the relevant target domain."[52]

The method is not a set of rigid rules but rather the principled and articulated decision to abandon some bad epistemic habits and methodological commitments that stand in the way of progress in comprehending the world and human behavior. It requires being resolutely engaged and pragmatically opportunistic in the use of whatever ideas and perspectives help when doing descriptive and explanatory work.

Consider again, in this connection, the question of whether every case of addiction is a brain defect or a neurological disease or disorder. A knee-jerk explanatory physicalism would say yes. However, the natural method, as Flanagan points out, does not try to shoehorn an understanding of addiction into the language of neuroscience. The natural method combines explanatory tools, research findings, and theoretically pregnant hypotheses from a host of fields—cognitive psychology, evolutionary biology, cultural anthropology, and neuroscience, to name four—in an effort to understand addiction. No such field or disciplinary perspective exists in isolation. Perspectives interact in relations of mutual influence and constraint. In effect, the natural method explanatorily *pluralizes* a disorder like addiction. It identifies complex social-environmental-cultural risk factors and stressors that impact proximate specific psychological processes and cognitive mechanisms that course through addictive behavior patterns.[53]

Conclusion

In 2009, the National Institute of Mental Health (NIMH), the world's most powerful neuroscience funding foundation, proposed an initiative to transform and reconfigure the current framework of psychiatric taxonomy and diagnosis—not just of addiction but of mental illness, period. The initiative was christened the Research Domain Criteria (RDoC) and is designed to offer a synoptic heuristic for grounding research into mental illnesses or disorders. One premise of the NIMH heuristic is the assumption that mental disorders are, in the words of the former director of the NIMH Thomas Insel and some of his colleagues, "disorders of brain circuits."[54]

The RDoC's assumption that mental disorders are neural disorders or defects, if read as a starting point for a classification of psychopathology, is at best an empirical gamble, at worst an explanatorily unwise methodological constraint. The claim goes beyond current evidence for the etiology of mental illnesses or disorders. The explanatory difference makers in the onset and persistence of a mental illness normally appear to be in important respects life-historical, situational, cultural, and psychological in description. Understanding the external context in which a person with a mental illness or disorder is embedded and how the individual perceives and experiences that context and its stresses and strains is crucial for understanding the emergence and persistence of a mental disorder.

Neuroscience already does, of course, help to contribute to the understanding of numerous mental disorders or illnesses. There is, for example, an established role for explanatory reference to dopamine release mechanisms in the study of effects of behavioral cues in those seeking addictive drugs.[55] Reference to mechanisms of release, however, does not necessarily require them to be described as defective or broken. Different classes of drugs exert different effects on those mechanisms and depend on an individual's past history and current life circumstances. In any case, the general methodological point is that brain science should not be allowed to *presumptively* reign over any and all other research frameworks for understanding addiction or, for that matter, for comprehending any mental illness. Rights of explanatory governance must be earned empirically.

Sometimes, defective circuits reign. Sometimes, circuits are not broken. Yet the doorbell rings when we wish that it did not. The squirrel, strain and stressor on human patience, always rings twice.

Notes

1. Numerous people offered helpful comments on drafts of this chapter or topics within it. These include Catherine Elgin, Nicholas Georgalis, Max Hocutt, Robert Kane, Peter Killeen, Joseph Lombardi, Robert McCauley, Robert Neville, and Serife Tekin, as well as various members of the audience when it was delivered via Skype

at the conference in Owen Flanagan's honor at Duke University, Durham, North Carolina, in September 2017. Special thanks are owed to G. Lynn Stephens. Indirect financial support came from The John Templeton Foundation.

2. Poland and Graham, *Addiction and Responsibility*; Levy, *Addiction and Self-Control*.

3. Flanagan, Identity and addiction, 886.

4. Ibid., 882.

5. "Addiction" of an unwilling type refers to an excessive and deleterious pattern of behavior, a form of behavior that is, as I describe elsewhere, "repeatedly engaged in despite its negative consequences, ... prefaced by impulses or motives that the addict wishes to control but finds difficult to consistently master, self-control or govern." See Graham, *The Disordered Mind*, 178. See also Stephens and Graham, Mental illness and the consciousness thesis; Stephens and Graham, An addictive lesson; Kennett, Just say no?

6. Flanagan, Identity and addiction, 886.

7. Flanagan, What is it like to be an addict?, 291.

8. Ibid.

9. Graham and Flanagan, *Psychiatry and the Brain*.

10. Keep in mind that natural selection or evolution is a "tinkerer, building new structures and systems by modifying existing structures, rather than by designing them from scratch." See Preuss, The argument from animals and humans in cognitive neuroscience, 496. So, if we ask, "Why isn't the brain resistant to addiction?," we must consider not only phenomena such as the capacity for learning on various reinforcement schedules but where the learning systems like those that are operative in reinforcement are derived from.

11. Leshner, Addiction is a brain disease, and it matters.

12. Gastfriend, Physician substance abuse and recovery, 1514.

13. Ross et al., *Midbrain Mutiny*.

14. The dopaminergic system is a reward learning system. The metaphor of hijacking is used to identify the (alleged) fact that the normal or healthy process of reward learning is neurologically defective or broken in a case of addiction.

15. Graham, *The Disordered Mind*; Stephens and Graham, Mental illness and the consciousness thesis.

16. Ibid.

17. See Pickard and Pearce, Addiction in context.

18. This is an interspecies linguistic echo of Jim Cain's 1934 novel *The Postman Always Rings Twice* and subsequent movie versions. See Cain, *The Postman Always Rings Twice.*

19. Neisser, The future of cognitive science.

20. West, *Theory of Addiction.*

21. The difference between good and bad bells (and other artifacts) depends on how we human beings conceive of their purposes. It depends on why we make them. If a manufacturer makes a bell to resist the pushes of squirrels and it fails to do so, this failure would be classified as a mechanical defect in the bell. The same is not true of brains. We don't make brains. We don't manufacture them to serve our purposes.

22. Profet, Pregnancy sickness.

23. Nesse and Williams, *Why We Get Sick.*

24. Stephens and Graham, Mental illness and the consciousness thesis, 395.

25. Flanagan, Phenomenal authority; West, *Theory of Addiction.*

26. Flanagan, Identity and addiction; Flanagan, The shame of addiction; Flanagan, Phenomenal authority; Flanagan and Graham, Truth and sanity; Graham, *The Disordered Mind*; Graham, Ordering disorder; Graham, Being a mental disorder; Graham and Flanagan, *Psychiatry and the Brain*; Tekin, Flanagan, and Graham, Against the drug cure model. See also Kendler, The dappled nature of causes of psychiatric illness; Lilienfeld, The research domain criteria; McCauley and Graham, *Hearing Voices and Other Matters of the Mind.*

Compare humans with rats. Sources of addiction in human beings are not as coarse grained and explanatorily single leveled as in rats. A rat may frantically press levers for more of a deleterious substance to which they (as it may be put) are addicted. But a rat's neural reward system is not accompanied in the cognitive economy of the animal by a conscious distinction between consuming substances that are prudent or healthy and those that are not. By contrast, a person may distinguish or comprehend a distinction between the dangers of addictive alcohol consumption and other forms of consumption that are of prudential benefit. See Graham, *The Disordered Mind*, 191–193. A person may make concerted efforts to refrain from imprudent patterns of consumption. One type of effort consists in removing oneself from physical closeness to drug-associated or other relevant stimuli, for instance. See Pickard and Pearce, Addiction in context, 174. Rats do not rise to such a conceptual and evaluative level of reflective analysis. A rat's preferences may change, of course, but not because they decide that their preferences should change or insofar as they make a deliberate effort to change.

27. Tillich, *Dynamics of Faith.*

28. Ibid., 105.

29. Neville, *God the Creator*, 214. Robert Neville was a philosophy professor of mine at Fordham. So were Robert Kane and Robert Mulvaney. I am grateful to each of them, not just as teachers but also as lifelong friends.

30. Ibid.

31. Bentall, *Doctoring the Mind*, 266.

32. Frankfurt, Rationality and the unthinkable, 190.

33. Nozick, *Philosophical Explanation*, 579.

34. Flanagan, *The Really Hard Problem*, 204.

35. Ross et al., *Midbrain Mutiny*, 194–195.

36. Flanagan, *The Really Hard Problem*, 199.

37. The moral engagement to which Flanagan refers is not conceptually monolithic. "Most of us ... do not have a blueprint for our whole lives," he aptly notes. But gaining certain goods of life, such as security, self-respect, social trust, and friendship, involves morally evaluating ourselves. It requires situating ourselves compassionately in moral-social space. See Flanagan, Identity and addiction, 2.

38. Graham, Identity and agency; Graham, *The Abraham Dilemma*; Garrett and Graham, At the empathetic center of our moral lives.

39. Frances, *Saving Normal*, 179–180.

40. Adopting an ethical and compassionate perspective raises questions about which values, norms, or moral principles are best or most defensible, questions that are addressed by Flanagan in various publications. See Flanagan, *The Really Hard Problem*; Flanagan, *The Geography of Morals*. Those questions are much too complex to explore in the space of the present chapter.

41. Flanagan, *The Really Hard Problem*, 218.

42. Cited in McKay and Whitehouse, Religion and morality, 450.

43. Boyer, Why is religion natural?, 29.

44. Bloom, *Just Babies*, 205.

45. Graham, Identity and agency; Graham, *The Abraham Dilemma*. In the Christian tradition, which is the tradition represented by Owen's and my collegiate alma mater, the demands of moral ideals are dramatically exemplified by the person of Christ. The Gospels tell different stories about Jesus's character, but a captivating image emerges so that numerous people become (as Owen describes them) advocates of Jesusism. ("Jesusism" is Owen's neologism, a linguistic sibling of "Buddhism.") "Late have I loved you, beauty so old and so new: late have I loved you," "You called

and cried out and shattered my deafness," and "You put to flight my blindness," Augustine (354–430) wrote in the *Confessions*. See Augustine, *Confessions*, bk. 10, 38.

46. Flanagan, *The Really Hard Problem*, 66; see also McCauley, *Why Religion Is Natural and Science Is Not.*

47. My own view of the warrant for and against theism or for being a theist is connected with the pragmatics of theistic belief. See Graham, *The Abraham Dilemma*.

48. To speak of the world's plural or many ways does not mean that we may arbitrarily choose how to describe or conceive of things. Pluralism, in the words of Nelson Goodman (1906–1998), "implies no relaxation of rigor." See Goodman, *Ways of Worldmaking*, 5. See also Goodman, The way the world is; McCauley, *Why Religion Is Natural and Science Is Not*; McCauley and Graham, *Hearing Voices and Other Matters of the Mind*.

49. Goodman, *Ways of Worldmaking*, 21.

50. Ibid.

51. Ibid.

52. Flanagan, *Dreaming Souls*, 17.

53. See also Kendler, The dappled nature of causes of psychiatric illness.

54. Insel et al., Research domain criteria, 749.

55. West, *Theory of Addiction*.

Bibliography

Augustine. *Confessions*. Translated by Henry Chadwick. Oxford: Oxford University Press, 1998.

Bentall, R. *Doctoring the Mind: Is Our Current Treatment of Mental Illness Really Any Good?* New York: New York University Press, 2009.

Bloom, Paul. *Just Babies: The Origins of Good and Evil*. New York: Crown, 2013.

Boyer, Pascal. Why is religion natural? *Skeptical Inquirer* 28, no. 2 (March/April 2004): 25–31.

Cain, Jim. *The Postman Always Rings Twice*. 1934. Reprint, New York: Penguin, 1989.

Flanagan, Owen. *Dreaming Souls: Sleep, Dreams, and the Evolution of the Conscious Mind*. Oxford: Oxford University Press, 2000.

Flanagan, Owen. *The Geography of Morals: Varieties of Moral Possibility*. Oxford: Oxford University Press, 2016.

Flanagan, Owen. Identity and addiction: What alcoholic memoirs teach. In *The Oxford Handbook of Philosophy and Psychiatry*, edited by K. W. M. Fulford, M. Davies, R. Gipps, G. Graham, J. Sadler, G. Stanghellini, and T. Thornton, 865–888. Oxford: Oxford University Press, 2013.

Flanagan, Owen. Phenomenal authority: The epistemic authority of Alcoholics Anonymous. In *Addiction and Self-Control: Perspectives from Philosophy, Psychology, and Neuroscience*, edited by N. Levy, 67–93. Oxford: Oxford University Press, 2013.

Flanagan, Owen. *The Really Hard Problem: Meaning in a Material World.* Cambridge, MA: MIT Press, 2007.

Flanagan, Owen. The shame of addiction. *Frontiers in Psychiatry* 4 (2013): 1–11.

Flanagan, Owen. What is it like to be an addict? In *Addiction and Responsibility*, edited by Jeffrey Poland and George Graham, 247–292. Cambridge, MA: MIT Press, 2011.

Flanagan, Owen, and Graham, George. Truth and sanity: Positive illusions, spiritual delusions, and metaphysical hallucinations. In *Extraordinary Science and Psychiatry: Response to the Crisis in Mental Health Research*, edited by Jeffrey Poland and Şerife Tekin, 293–314. Cambridge, MA: MIT Press, 2017.

Frances, A. *Saving Normal: An Insider's Revolt against Out-of-Control Psychiatric Diagnosis, DSM-5, Big Pharma, and the Medicalization of Ordinary Life*. New York: HarperCollins, 2013.

Frankfurt, H. Rationality and the unthinkable. In *The Importance of What We Care About*, edited by H. Frankfurt, 177–190. Cambridge: Cambridge University Press, 1988.

Garrett, R., and Graham, George. At the empathetic center of our moral lives. In *Empathy and Morality*, edited by H. Maibom, 122–137. New York: Oxford University Press, 2014.

Gastfriend, D. R. Physician substance abuse and recovery: What does it mean for physicians—and everyone else? *Journal of the American Medical Association* 293, no. 12 (2005): 1513–1514.

Goodman, Nelson. *Ways of Worldmaking*. Indianapolis, IN: Hackett, 1978.

Goodman, Nelson. The way the world is. *Review of Metaphysics* 14 (1960): 48–56.

Graham, George. *The Abraham Dilemma: A Divine Delusion*. Oxford: Oxford University Press, 2015.

Graham, George. Being a mental disorder. In *Classifying Psychopathology: Mental Kinds and Natural Kinds*, edited by H. Kincaid and J. Sullivan, 123–143. Cambridge, MA: MIT Press, 2014.

Graham, George. *The Disordered Mind: An Introduction to Philosophy of Mind and Mental Illness*. 2nd ed. Oxford: Routledge, 2013.

Graham, George. Identity and agency: Conceptual lessons for the psychiatric ethics of patient care. In *The Oxford Handbook of Psychiatric Ethics,* edited by J. Sadler, K. Fulford, and C. van Staden, 372–386. Oxford: Oxford University Press, 2015.

Graham, George. Ordering disorder: Mental disorder, brain disorder, and therapeutic intervention. In *The Oxford Handbook of Philosophy and Psychiatry,* edited by K. Fulford, M. Davies, R. Gipps, G. Graham, J. Sadler, G. Stranghellini, and T. Thornton, 512–530. Oxford: Oxford University Press, 2013.

Graham, George, and Flanagan, Owen. "Psychiatry and the Brain." *OUPblog,* August 1, 2013. http://blog.oup.com/2013/08/psychiatry-brain-dsm-5-rdoc/.

Insel, Thomas, Cuthbert, Bruce, Garvey, Marjorie, Heinssen, Robert, Pine, Daniel S., Quinn, Kevin, Sanislow, Charles, and Wang, Philip. Research domain criteria (RDoC): Toward a new classification framework for research on mental disorders. *American Journal of Psychiatry* 167 (2010): 748–751.

Kendler, K. S. The dappled nature of causes of psychiatric illness. *Molecular Psychiatry* 17, no. 4 (2012): 377–388.

Kennett, J. Just say no? Addiction and elements of self-control. In *Addiction and Self-Control: Perspectives from Philosophy, Psychology, and Neuroscience,* edited by N. Levy, 144–164. Oxford: Oxford University Press, 2013.

Leshner, A. Addiction is a brain disease, and it matters. *Science* 278 (1997): 45–47.

Levy, N. *Addiction and Self-Control: Perspectives from Philosophy, Psychology, and Neuroscience.* Oxford: Oxford University Press, 2013.

Lilienfeld, Scott O. The research domain criteria (RDoC): An analysis of methodological and conceptual challenges. *Behaviour Research and Therapy* 62 (2014): 129–139.

McCauley, Robert. *Why Religion Is Natural and Science Is Not.* Oxford: Oxford University Press, 2011.

McCauley, Robert, and Graham, George. *Hearing Voices and Other Matters of the Mind: What Mental Abnormalities Can Teach Us about Religions.* New York: Oxford University Press, 2020.

McKay, R., and Whitehouse, H. Religion and morality. *Psychological Bulletin* 141, no. 2 (2015): 447–473.

Neisser, U. The future of cognitive science: An ecological analysis. In *The Future of the Cognitive Revolution,* edited by D. Johnson and C. Erneling, 247–273. New York: Oxford University Press, 1997.

Nesse, R., and Williams, G. *Why We Get Sick: The New Science of Darwinian Medicine.* New York: Oxford University Press, 1996.

Neville, R. *God the Creator: On the Transcendence and Presence of God*. Chicago: University of Chicago Press, 1968.

Nozick, R. *Philosophical Explanations*. Cambridge, MA: Harvard University Press, 1981.

Pickard, H., and Pearce, S. Addiction in context: Philosophical lessons from a personality disorder clinic. In *Addiction and Self-Control: Perspectives from Philosophy, Psychology, and Neuroscience*, edited by N. Levy, 165–189. Oxford: Oxford University Press, 2013.

Poland, Jeffrey, and Graham, George. *Addiction and Responsibility*. Cambridge, MA: MIT Press, 2011.

Preuss, T. The argument from animals and humans in cognitive neuroscience. In *Cognitive Neuroscience: A Reader*, edited by M. Gazzaniga, 483–502. Malden, MA: Blackwell, 2000.

Profet, M. Pregnancy sickness: A deterrent to maternal ingestion of teratogens. In *The Adapted Mind: Evolutionary Psychology and the Generation of Culture*, edited by J. Barkow, L. Cosmides, and J. Tooby, 203–220. Oxford: Oxford University Press, 1992.

Ross, D., Sharp, C., Vuchinich, R., and Spurrett, D. *Midbrain Mutiny: The Picoeconomics and Neuroeconomics of Disordered Gambling*. Cambridge, MA: MIT Press, 2008.

Stephens, G. L., and Graham, George. An addictive lesson: A case study in psychiatry as cognitive neuroscience. In *Psychiatry as Cognitive Neuroscience*, edited by M. Broome and L. Bortolotti, 203–220. Oxford: Oxford University Press, 2009.

Stephens, G. L., and Graham, George. Mental illness and the consciousness thesis. In *The Neuropsychology of Mental Illness*, edited by S. Wood, N. Allen, and C. Pantelis, 390–398. Cambridge: Cambridge University Press, 2009.

Tekin, S., Flanagan, O., and Graham, G. Against the drug cure model: Addiction, identity and pharmaceuticals. In *Philosophical Issues in Pharmaceuticals*, edited by H. Do, 221–236. Dordrecht: Springer, 2017.

Tillich, P. *Dynamics of Faith*. New York: Harper and Brothers, 1957.

West, R. *Theory of Addiction*. Oxford: Blackwell, 2006.

5 On the Use of the Notion of Narrative in Ethics and Psychology

Galen Strawson

> Life is not susceptible perhaps to the treatment we give it when we try to tell it.
> —Virginia Woolf, *The Waves*

> A man's road back to himself is a return from his spiritual exile, for that is what a personal history amounts to—exile.
> —Saul Bellow, *The Actual*

Introduction

The theoretical use of the word "narrative" is reasonably clear in its home domain, the study of the arts. Things are much less clear when we consider its theoretical use in other domains. I'm going to argue that almost all claims made using "narrative" in ethics and psychology are either false or truistic, trivially true, and so *empty* in a sense I'll explain.

I'll start with a *definition*, and a remark about *memory*. Then I'll provide some *data*—examples of the use of terms like "narrative" or "story in" ethics and psychology. Then I'll set down four *theses* that are very often in play when it comes to applications of the term in ethics and psychology, and comment on them. Then I'll list eight *platitudes* about human life in order to argue that the central narrativist thesis is trivially true if true at all, and otherwise false. I'll end with a few further thoughts.

Definition

I'm going to use the words "narrative" and "narrativity" to denote a psychological trait when applied to people. Here is an initial definition. If one is *narrative*,

> one experiences or conceives of one's life, one's existence, in a narrative way, as having the form of a story, or perhaps a collection of stories, and—in some manner—one lives in and experiences oneself through this conception.

Many of us have a sense that this could be a plausible description of a human being, even if it seems considerably less clear when we begin to reflect on what exactly it says. In the same way, I think, we can grasp the idea that someone might be *nonnarrative* and not naturally experience themselves or their life in this way.

I think, furthermore, that some people are *antinarrative*. They don't simply lack any natural tendency to think of themselves and their life in a narrative or storylike way. Their whole way of thinking of themselves and their life resists narrative form or narrative patterning. Their memories don't operate in a narrative way.

I think this is not uncommon—I think I am that way myself. I feel an affinity with Montaigne, who "seems to have been allergic to sequential order and suspicious of biographical coherence."[1] When Montaigne wrote that "nothing is so contrary to my style [*mon stile*] as extended narration [*une narration estendue*]. I have to break off so often from shortness of wind that neither the structure [*composition*] of my works nor their development [*explication*] is worth anything at all,"[2] he had more than his writing style in mind.

According to this proposal, narrativity, or the lack of it, is in the first instance a natural psychological disposition, even if it's open to cultivation. It's a natural and ethically important dimension of human psychological difference, whatever the possible effects of training or cultural influence, profession or current preoccupation, mood, health, or age.

Memory

We human beings differ profoundly in the way we experience and handle memory. We differ in the amount of time we spend remembering things, in what kinds of things we remember, and in how much of our lives we remember. We differ astonishingly in the accuracy of our memories. Taken as a whole, these are among the largest and deepest differences between us. I think that the extent of the variation deserves far more study from psychologists.[3]

Memory differences are linked to deep differences in intellectual and ethical style. Philosophers often ignore them, both in moral psychology and in ethical theory generally. They tend to presuppose that people are fundamentally similar with respect to memory. It's not that they don't acknowledge the differences when they're described. It's rather that they don't sufficiently recognize their importance.

I doubt that there is any more important difference between human beings when it comes to moral psychology. I think many confident generalizations about human beings may fail for reasons connected to differences in memory and, more generally, that statistical generalizations tend to obscure a vast dimension of variation in cases where the variations are precisely what matter most.[4]

I remember little of my own past life. My father also had little memory of his own past, and almost never thought about it. Elie Wiesel is very different; for him, "memory is a passion no less powerful or pervasive than love."[5] Oskar Frisch is different again: "I have always lived very much in the present, remembering only what seemed to be worth retelling."[6]

The nameless founder of the Mnemosyne Institute in Saul Bellow's *The Bellarosa Connection* is more like Wiesel: "As I used to say to clients, 'Memory is life.'"[7] I'm with Montaigne when, remarking that he is "better at friendship than at anything else," he points out that friendship doesn't depend on memory: "There is nobody less suited than I am to start talking about memory. I can find hardly a trace of it in myself; I doubt if there is any other memory in the world as grotesquely faulty as mine is!"[8] It's interesting to think about the long centuries when there were no photographs.

Gilbert Ryle thinks that "being good at recalling is being good at presenting.... It is a narrative skill."[9] I doubt this, except insofar as it's a tautology (being good at recalling things, in such a way as to be able to present them in a connected manner agreeable to listeners, is a narrative skill). For the rest, I think it's the reverse of the truth. Sue Halpern says that "memory ... is an expert story teller. If there is a context, it fills in the blanks. If there are characters, it fleshes them out."[10] It appears to follow that being good at recalling is being bad at recalling. In my own case, memory almost always says one thing: "I can't remember."[11]

Halpern also says that "we are made of memories.... Memory constitutes personality; it both holds and defines our selves."[12] This would be worrying if true, if memory confabulates as she says. But it isn't true. Personality is largely set by genetic inheritance and early upbringing, and while one's subsequent experiences continue to shape one's personality in certain ways (shaping is far weaker than constituting), one certainly doesn't have to remember the experiences for this to be so, and it may be better if one doesn't.[13]

These are perhaps obvious points, and not particularly interesting as such. What is interesting, to people like me, is that thoughtful people can think that sweeping claims like Halpern's are true. We wonder what it is like to be them. But the present point is simply that the differences between them and us provide further dramatic confirmation for the view that radical differences in memory are among the deepest differences between human beings. They're among the most important differences when we consider fundamentals of character, personality, temperament, or moral style. This has become vivid to me in a way in which it never would have if I hadn't written about this topic and encountered a wide variety of different reactions.

Data

I've offered an initial definition of narrativity, but the question remains. What is it, exactly, to experience oneself or one's life, or major parts of one's life, in a narrative way? It's a large question and it has no generally accepted answer. But there's widespread agreement that narrativity exists. Here are some data, gathered from some notable sources.

According to the poet and essayist Charles Simic, "We make sense of our lives not just by ferreting out the facts, but by turning them into stories, so we can bring past events to life."[14] The clinical neurologist Oliver Sacks says that "each of us constructs and lives a 'narrative' ... this narrative *is* us, our identities."[15] According to the psychologist Jerome Bruner, "Self is a perpetually rewritten story."[16] According to the philosopher Alasdair MacIntyre, "We all live out narratives in our lives and ... understand our own lives in terms of the narratives that we live out."[17]

Another philosopher, Charles Taylor, thinks that "we must inescapably understand our lives in narrative form."[18] The literary critic Barbara Hardy claims that "we dream in narrative, day-dream in narrative, remember, anticipate, hope, despair, believe, doubt, plan, revise, criticize, construct, gossip, learn, hate and love by narrative."[19] The philosopher Marya Schechtman once wrote that "a person creates his identity by forming an autobiographical narrative—a story of his life."[20] Clinical psychologist Ruthellen

Josselson, psychologist Amia Lieblich, and social psychologist Dan McAdams say, in a joint paper, that "we are all storytellers, and we are the stories we tell."[21] The philosopher David Velleman holds that "we invent ourselves [and] we really are the characters whom we invent."[22] Owen Flanagan has spoken of the "self of narrative construction"[23] and "the story we tell ourselves to understand ourselves for who we are."[24] Philip Pullman writes that:

> I find it so natural and instinctive to shape my own experience in terms of story that I find it hard to understand that other people don't. Apart from my own "story," ... for as long as I can remember I've been aware of part of my mind observing events in terms of how I'd use them in a story: how this would lead to that, and be resolved by such-and-such, and point to a hidden connection, which in turn would ... and so on. I really thought that everyone did that.[25]

Novalis may approve, although it's not clear exactly how to interpret his claim that "life must not be a novel that is given to us, but one that is made by us."[26] Jean-Paul Sartre, however, does not approve. After observing that "a man is always a teller of stories, he lives surrounded by his own stories and those of other people, he sees everything that happens to him *in terms of* these stories and he tries to live his life as if he were recounting it,"[27] he famously adds: "Mais il faut choisir: vivre ou raconter" ("But one must choose: live or tell stories"). In his view, it seems, narrativity is antilife—inauthenticity.[28] But many today—the *narrativists*, those who endorse the psychological narrativity thesis—seem to think that to live one's life (*vivre*) and to narrate it (*raconter*) are the same, or at least that the first essentially involves the second. The novelist Julian Barnes writes of "the narratives we turn our lives into,"[29] but he is as suspicious of this activity as Sartre is. The philosopher Iris Murdoch agrees, saying that "man is a creature who makes pictures of himself, and then comes to resemble the picture."[30] In her view, this is a large part of the reason why "the self, the place where we live, is a place of illusion."[31] Flanagan has recently expressed strong reservations about the dominant narrativist approach, arguing for the importance of recognizing a person's "non-narrative, non-forensic *dasein*" (one's *dasein* is, roughly, one's personal being).

These quotations give the general idea. Some of them combine the claim that one experiences *one's life* in a narrative way with the claim that one conceives of *oneself* in a narrative way, and make no clear distinction between the two things. This is common, so I've built it into the definition of narrativity.

The Psychological Narrativity Thesis

Given our initial definition of narrativity, one can distinguish at least four different attitudes toward it. All the people I've just quoted endorse what I call

(1) the *psychological narrativity thesis,*

according to which all normal people (I use "normal" in a very wide sense) are narrative. If one divides it up as

(i) we all *conceive* ourselves and our lives in a narrative way,

(ii) we all *make sense* of ourselves and our lives in a narrative way,

(iii) we all *live in and through* a narrative conception of ourselves and our lives,

one can weaken (ii) to

(ii) we all try to make sense of ourselves and our lives in a narrative way *insofar as we try to make sense of these things at all,*

but I think it's still false in this version—or else trivially true. I'll try to say why, after considering three further theses.

The Narrative Self-Constitution Thesis

To (1), the psychological narrativity thesis, we can add

(2) the *narrative self-constitution thesis,*

which endorses and extends the psychological narrativity thesis by adding

(iv) all normal people *constitute their identity* as persons or selves by virtue of being narrative.

"We are the stories we tell."[32] "Each of us constructs and lives a 'narrative.' ... This narrative *is* us, our identities."[33] "We *become* the autobiographical narratives by which we 'tell about' our lives."[34] According to Barbara Hardy, "We tell stories in order to change, remaking the past in a constant and not always barren *esprit d'escalier.*"[35]

I think Hardy may speak for many, but she doesn't speak for all. I don't doubt that this is a sincere self-report, but I for one have no idea what she is talking about. Some of us find the past impressively resistant to change. And if "*narrative identity* is the internalized and evolving story a person invents [n.b.] to explain how he or she has become the person he or she is becoming," as McAdams claims in a recent paper,[36] then I don't have any narrative identity at all. In his account, I'm not sure I qualify as human.

Some may think that the narrative self-constitution thesis as just formulated doesn't really add anything to the psychological narrativity thesis summarized in (i)–(iii). I disagree. It seems to me that (iv), the claim that we constitute our identity in a narrative way, is a very strong claim that goes beyond all of (i)–(iii) and is worth listing separately.

I think (iv) is false. I don't think that we have any significant say in the matter of our identity. There is, however, a deep difference between the ways philosophers and psychologists understand the notion of a person's identity. Psychologists usually mean what Flanagan calls one's "self-represented identity" or "SRI," one's self-*image*; philosophers usually mean something quite different, which Flanagan calls one's "actual full identity" or "AFI."[37] One's AFI is the actual nature of one's character, one's psychological deep essence, "God's description of [one's] identity."[38]

I take one's identity to be one's AFI. That is one reason why I have no doubt that the narrative self-constitution thesis is false. I think, in fact, that it is no less false if one takes one's identity to be one's SRI, but I'm happy to allow (at least at this stage of the argument) that there may be some people who really are narrative types and who really do partly constitute their SRI by virtue of being narrative.

Philosophers and psychologists may question—even scorn—the notion that there is any such thing as the psychological deep essence of a person. So it goes. No one who denies the validity of this notion has a clue about human beings.[39]

The Ethical Narrativity Thesis

Many of those who endorse the descriptive, empirical-psychological narrativity thesis also endorse a related *normative* thesis,

(3) the *ethical narrativity thesis,*

according to which

(v) one ought to have a narrative outlook on one's life. One ought to think about oneself and one's existence in a narrative way and in some manner live in and through this conception. It's essential to a good life, essential to living well, essential, in fact, to true or full *personhood.*

Sartre thinks this is completely wrong: "One must choose: live or tell stories." But it's a very popular view today.

We can illustrate it by supposing that Socrates is right when he says that the unexamined life is not a full human life, not a life for a human being. We can suppose he's right that we ought to be reflective about our lives in some way or other. I'm not at all sure that this is a universal human truth—"an unexamined life can be virtuous," as Iris Murdoch points out[40]—but let's assume for now that it is true. We may then take the ethical narrativity thesis to be the thesis that we ought to reflect about our lives in a particular way: in the narrative way, in the narrative framework.[41]

Paul Ricoeur thinks something like this. He claims that "self-understanding is an interpretation" and holds that "interpretation of the self ... finds in narrative ... a privileged form of mediation." So far this sounds reasonable enough. But he goes on to say that this mediation (in or by narrative) "borrows from history as well as from fiction, making a life story a fictional history or, if one prefers, a historical fiction."[42] He finds no difficulty in the idea that all self-understanding involves fiction, but it seems to me very alarming. If self-narration does essentially involve fiction, then I hope that the ethical narrativity thesis isn't true, and I have a further reason to believe it isn't true.

This is another deep human difference. Some seem easy with the idea that their sense of their life is suffused with fiction and invention; others can't imagine wanting anything other than the truth. I'm in the second camp, helplessly so. Joan Didion, in her essay "On Keeping a Notebook," is in the first:

> I tell what some would call lies. "That's simply not true," the members of my family frequently tell me when they come up against my memory of a shared event. "The party was *not* for you, the spider was not a black widow, *it wasn't that way at all.*" Very likely they are right, for not only have I always had trouble distinguishing between what happened and what merely might have happened, but I remain unconvinced that the distinction, for my purposes, matters.[43]

As Martin Amis observed in his review of Didion's collection *The White Album* (which opens with her well-known claim that "we tell ourselves stories in order to live"), "'how she feels' has become, for the time being, how it is."[44]

I don't know what to think, other than that this must very often be an indicator of abuse and trauma. Perhaps Shelley Taylor and Jonathon Brown, in their famous paper, are right that "positive illusions" make people's lives go better;[45] I'm sticking with Descartes, in his letter to Princess Elisabeth of Bohemia:

> Is it better to be cheerful and content, imagining the goods one possesses to be greater and more valuable than they are, and not knowing or caring to consider those one lacks; or is it better to have more consideration and knowledge, so as to know the just value of both, and thus grow sad? ... I must conclude that it is better to be less cheerful and possess more knowledge.... It is not always the most cheerful person who has the most satisfied mind.[46]

There's a further point. I've allowed supporters of the ethical narrativity thesis to cite Socrates in support of their position, but it doesn't follow, from his claim that one ought to reflect on life ("the unexamined life is not a full human life"), that one ought to reflect specifically and directly on one's own life; still less, that one ought to do so in a specifically "narrative" fashion. One can learn a very great deal from reading novels,[47] and it may in general be far more instructive and more salutary to think more about others than about oneself. Insofar as it is helpful to think about oneself, it may be better to think about oneself in a nonnarrative way. So even if Socrates' famous remark is true for almost all of us, it offers no direct support to the ethical narrativity thesis.

The Ethical Narrative Self-Constitution Thesis

The ethical narrativity thesis is very widely endorsed today, although there are of course dissenters. It's an article of faith for many philosophers, psychologists, therapists, and authors of "self-help" books, at least in the West. It's also widely accepted in the general population, and it's often coupled with a normative version of the narrative self-constitution thesis,

(4) the *ethical narrative self-constitution thesis*,

according to which

(vi) one's identity as a person or self ought to be constituted by (or perhaps in or through) one's narrative of one's life.

We have, then, a major if imperfect consensus: (i) we *conceive* of ourselves and our lives in a narrative way; (ii) we *make sense* of ourselves and our lives in a narrative way (insofar as we try to make sense of these things at all); (iii) we *live in and through* a narrative conception of ourselves and our lives, whether or not (iv) we *constitute our identity* in a narrative way. And we ought to do these things: (v) we ought to conceive and make sense of things in this way and live in this way; and (vi) we ought to constitute our identity in this way. The consensus contains many variations, but it's plainly a consensus of some kind, and there may be a way of understanding (i)–(iii), the basic psychological narrativity thesis, given which it's correct—even if the same can't be said of (iv), the narrative self-constitution thesis, or (v) and (vi), the two ethical or normative theses.

If this is so—if there's a sense in which the psychological narrativity thesis is correct—then of course I endorse it. Truth is good, obvious or not. I am, however, going to assume that the psychological narrativity thesis is meant to be unobviously true in at least the following sense. It's meant to add something new—importantly new—to our existing general picture of what it is to be a human being and live a human life. It's meant to add something we didn't clearly realize before the thesis was formulated.

Triviality 1

I'll say that a thesis introducing a term "F" is *trivial*, trivially *true*, if it follows unsurprisingly from widely accepted basic facts that are equally well expressed in other terms (i.e., without using "F"). A thesis about narrative or narrativity is trivially true if it follows from widely agreed basic facts about human life that are naturally and fully expressed without using terms like "narrative," "story," or "tale." It's trivial in the sense that it adds nothing to what we already standardly take to be true when we think about the matter in other terms. It merely reexpresses it in a different vocabulary. It's reasonable to ask whether the psychological narrativity thesis is trivial in this sense—insofar as it's true at all—because it's meant to be importantly new. It's meant to bring into view something we didn't already know, or hadn't explicitly formulated, and now see to be true. It's meant to add something true and fundamental to our conception of what a person—or at least a human being—is.

"Not so fast. A proposition *q* may be unobviously entailed by some other proposition or set of propositions *p* in such a way that it's natural to feel that to articulate *q* is to add something genuinely new to the picture of how things are that is expressed by *p* even though *q* is in fact already entailed by *p*. In such cases, it isn't adequate (it isn't fair) to say that *q* simply reexpresses in a different vocabulary, or in a different way, what *p* is already understood to express."

True (I reply). *q* may even be pretty obviously entailed by *p* and still appear to add something genuinely new to our picture of how things are. The appearance will be deceptive, strictly speaking, but even if *q* involves little more than a change of word, relative to *p*, it can be theoretically valuable if it helps to revivify or cast a sharper light on a familiar idea—even if it's trivial in the present sense. If so, triviality in the present sense may not be a great fault.[48]

True again. But the charge of (relative) triviality is worth investigating. I'll call a claim or thesis that is trivial in this sense "empty." It won't be empty in the sense of being contentless; it will be both meaningful and true. It'll be empty in the sense that it contributes nothing to common knowledge but simply rewords it.

So charges of emptiness will have the following form: [i] It's very widely agreed—and true—that *x* is *F*, that *x* is an *F*, or that *x*s are *F*. [ii] What is meant by "*x* is *G*," "*x* is a *G*," or "*x*s are *G*" adds nothing—nothing true—to what is already expressed by "*x* is *F*," "*x* is an *F*," or "*x*s are *F*." So [iii] "*x* is *G*," "*x* is a *G*," or "*x*s are *G*" is empty in the current sense (trivially true). In the present case, the candidate replacement for "*G*" is "narrative." The relevant claims are "human life is narrative," "human life is a narrative," "all normal human beings are narrative." "*F*" summarizes a body of platitudes about human life that I'm about to set out.

We can knock out the claim that "a human life is a narrative" straightaway. A narrative is essentially an account or representation of something, and a human life isn't in itself, let alone essentially, a representation of anything. It isn't a representation even if it involves a lot of representing, including self-representing, and even if that self-representing has a considerable impact on its character.[49] As William Blattner has observed, "We are not texts. Our histories are not narratives. Life is not literature."[50] We can treat the claim that life is a narrative as a metaphor, and even accept it as such, but we should then put it aside, because it introduces confusion into a debate that includes many very careless thinkers.

Eight Platitudes

Here are some simple and overlapping truths about human life that are very widely accepted. I list them here because many of them have been invoked in support of various narrativist claims. I'll call the first point "the *Locke* point," *Locke* for short.

[1] **Locke** All ordinary people fulfill a fundamental condition on what it is to be a person, a condition that is explicitly stated in Locke's famous definition of a person. They can "consider [themselves] as [themselves], the same thinking thing, in different times and places."[51] In other words, they are [i] fully self-conscious beings, on the terms of the standard definition of self-consciousness in analytic philosophy, and [ii] their self-consciousness has a significant temporal extent.[52]

The *Locke* point is well known to every remotely normal adult human being on the planet. (One doesn't have to believe in a persisting self in order to accept the core sense in which [i] and [ii] are true of ordinary human beings.) Clearly, if the truth of the *Locke* point were sufficient for the truth of the psychological narrativity thesis, as Anthony Rudd, for example, seems to suppose,[53] then the psychological narrativity thesis would be empty in the current sense.

Next, we have the *Self-History* point:

[2] **Self-History** Almost all ordinary people have a reasonably good grasp of their own history—basic facts about their own life. They possess a more or less adequate "self-history," a basic "self-chronicle," where the word "history" signifies basic factual accuracy.

I say "almost all" because there are people whose memory of their own past is so garbled that they can no longer be said to have an adequate self-history, although they're not classified as mentally unwell.[54] Using the term "self-biography" to mean a person's set of beliefs about their life, we can say that such people have a self-biography even if they don't have a self-history, because their self-biography is so dramatically at odds with their actual history. The present point is simply that if either [1] or [2] is sufficient for narrativity, then the psychological narrativity thesis is empty.

The third point is the *Timeline* point:

[3] **Timeline** All ordinary people know how old they are, more or less precisely, and they know where they are on the timeline of human life—however much or little it matters to them.

This is plainly a very important fact about human life.[55] The present point is that if [1], [2], and [3] are sufficient for narrativity, the psychological narrativity thesis is empty.

The fourth point scoops up our knowledge of many other basic facts about the world, including many more facts about human life. I'll call it *Knowledge 101*:

[4] **Knowledge 101** In our world, one thing leads to another in a highly regular fashion. We are constantly and vividly aware of this.[56] Our lives are set among, and involve, vast series of complexly causally connected happenings in time—coherent, sequentially structured, developmental processes, shorter or longer—nearly all of which have relatively well-defined beginnings and ends and typical intermediate stages, and are clearly understood by us to have this form: sleeping and waking, working and resting, sowing and reaping, aging, teething, courting, changing height, changing weight, changing levels of debt and pain, bodily cycles, illnesses, days, seasons, journeys, hair growth, hair loss, raising children, acquiring and losing skills, qualifications, and pension funds. We're embedded and engaged in thousands of intelligible, relatively long-term orderings of this sort, from birth to death. We're no less embedded and engaged in hundreds of thousands of shorter-term orderings, from breath to breath, dressing and undressing, washing dishes, singing, cooking, bathing, eating, digesting, excreting. We have to follow a series of steps to make coffee, we have to know what they are, and we're more or less sharply aware of them in taking them. Many thousands of characteristically human things involve such sequences and patterns. We all know this.

This is another obvious and fundamental fact about human life. The present point is that if [1], [2], [3], and [4] are sufficient for narrativity, the psychological narrativity thesis is empty.

The fifth point, *Explanation*, expands an aspect of *Knowledge 101*:

[5] **Explanation** Causal knowledge, causal explanation, is fundamental to our lives. Causal explanation almost invariably involves temporal order.[57] Causation is the "because something is, something else must be" relation.[58] It's an understanding of how one thing leads to another or connects to another.

This is obvious. The present point is that if [1], [2], [3], [4], and [5] are sufficient for narrativity, the psychological narrativity thesis is empty.

We're sometimes said to be narrative because narrative explanation is fundamental to our lives, but "narrative explanation" often turns out to be nothing other than explanation—causal explanation: this happened because that happened. Causal explanation almost invariably involves reference to temporal order, as just remarked, and sometimes, it seems, mere awareness of temporally ordered connection is taken to be enough for narrativity. At the limit, any description of anything to do with one's life that involves change is liable to be called a "narrative."

The sixth point, *Psychology*, also expands an aspect of *Knowledge 101*:

> [6] **Psychology** A vast number of significant and interesting facts about human beings' lives are facts about their psychological states—their hopes, fears, beliefs, desires, goals, memories, intentions, and so on. We all know this. Vast numbers of these psychological states involve explicit representations of connections between past and future states of the world and, in particular, past and future parts of one's own life. We all know this. Vast numbers of these psychological states are crucially involved in the vast numbers of regular sequences and processes that structure a person's life. We all know this.

This is obvious. The present point is that if the truth of *Psychology* is sufficient for the truth of the psychological narrativity thesis, with or without [1]–[5], then the psychological narrativity thesis is empty in the current sense.

The seventh point, *Action*, further expands *Knowledge 101*, *Psychology*, and *Explanation*:

> [7] **Action** Vast numbers of the sequences and processes mentioned in the description of the ordered temporal complexity of our lives in [4] play out as they do because of psychological states mentioned in [6]: wants, needs, intentions, likes and dislikes, goals, dreams, hopes, fears, suspicions, superstitions, and so on. Vast numbers of them also involve intentional action on our part, and almost all intentional action involves some anticipation, thinking ahead, planning, knowledge of steps to be taken, calculation of possible consequences, *what if?* thinking. It involves causal-temporal thinking, causal-temporal-psychological thinking: thinking that has causal matters and temporal matters and psychological matters as part of its content. We engage in such thinking all the time. And again, and of course, we know this.

This is again obvious, and the present point is that if *Action* (± [1]–[6]) is enough to make us narrative, the psychological narrativity thesis is empty. If any partial or total conjunction or disjunction of [1]–[7] is sufficient for the truth of the psychological narrativity thesis, then the thesis is empty in the present sense. If having goals is enough to make us narrative, the psychological narrativity thesis is empty. It must hope to amount to something more than this.[59]

Suppose someone says that the term "narrative" explanation doesn't apply to all causal explanations but only to explanations of intentional actions by reference to psychological states, and then says that we're essentially narrative because we're committed to narrative explanation in this sense. The charge of emptiness remains. "Why did she turn on the light? Because she wanted to see across the room." If this is already a narrative explanation, and if our commitment to the validity of such explanations is a key part of the psychological narrativity thesis, then the thesis is, so far, empty. The same goes when we explain a person's current disposition by reference to their past experiences. If the explanation is true, it's a straightforward causal explanation. It's exactly like the explanation of a bridge's collapse by reference to its having been weakened by a past impact.

Schechtman quotes MacIntyre, when characterizing narrative explanation: "To identify an occurrence as an action ... is to 'identify it under a type of description which enables us to see that occurrence as flowing intelligibly from a human agent's intentions, motives, passions, and purposes' (MacIntyre), *that is*, to tell a narrative that explains it."[60]

I've italicized the words "that is" because they produce the emptiness. So too when Schechtman writes that "for Taylor, as for MacIntyre, the intelligibility of action requires that our lives be narrative in form."[61] If the mere intelligibility of actions (their intelligibility by reference to reasons, motives, or causes for action) suffices for the claim that our lives are narrative in form, then our lives are certainly narrative in form, and so are the lives of vast numbers of animals. But the claim that they're narrative in form is empty in the present sense.

The point applies equally to Rudd's position. He quotes my remark that "if someone says, as some do, that making coffee is a narrative that involves narrativity, because you have to think ahead, do things in the right order, and so on, and that everyday life involves many such narratives, then I take it the claim is trivial,"[62] and comments that "what Strawson really seems to be doing is conceding (however grudgingly) that episodes like my coffee-making *are* teleological [i.e., done for a reason, done with an end in view] and *therefore* narrative."[63] This is a pure example of emptiness: all intentional action is narrative.[64]

Rudd goes on to say that narrativists "claim that we understand ourselves narratively insofar as we experience ourselves temporally."[65] But if that's all it takes for narrativity, then [1] alone is sufficient for narrativity, and the thesis is once again empty—trivially true.[66] We can add an eighth platitude:

[8] **Temporality** We all experience ourselves temporally simply in living from moment to moment as we do, when making coffee, remembering one thing, anticipating another. We experience ourselves temporally even when we're absorbed in what we're doing, living—as we say—wholly in the moment.

This adds little to the *Locke* and *Action* points, but it is perhaps worth listing separately.

Rudd also says that

> we may see a great constancy and consistency in someone's behaviour ... but it is only in a temporally extended pattern that one can see constancy. If character is essentially what expresses itself in and through someone's mental and physical states, then it is essentially something which unfolds in time. And this unfolding is something that can only be captured by narrative.[67]

But it isn't true that character [1] can only be captured in narrative. Nor is it true that character [2] essentially involves "unfolding in time." Nor is it true that [1] is true because [2] is true. One can give a brilliant description of character without any narrative description in Rudd's sense, and indeed without any description of temporally extended goings-on. It may be added (insofar as the notion of character is in good standing) that a person's character is wholly in place at any given time, whether it is known or not, and whether or not it changes over time.

But none of this matters at present. Nor does it matter that Rudd's conditional claim (if character essentially expresses itself in someone's mental and physical states, then it is essentially something that unfolds in time) is a non sequitur—an illegitimate slide from an epistemological claim to a metaphysical claim. The present problem is that this line of thought is another route to triviality.

Caesura

I don't think the psychological narrativity thesis is empty in the current sense, and I don't think that most of those who assert it take it to be empty in this sense, although they often advert to points like [1]–[7]—and even

[8]—when pushed to defend it.[68] I know [1]–[8] are true, and I fit their joint description of an ordinary human being, along with almost all adult human beings. But—as I understand the term "narrative"—I'm not in the least bit narrative (I'm certainly utterly unlike Philip Pullman and Joan Didion). So, I don't think the psychological narrativity thesis is empty, or in other words trivially true. I think it's false.

Triviality 2

We have eight overlapping points: *Locke, Timeline, Self-History, Knowledge 101, Explanation, Psychology, Action, Temporality*. As remarked, narrativists may feel that a proper acknowledgment of one or more of these points secures their case. But if it does, narrativism is empty in the current sense. For on this view, when we've gone over the well-known and obvious points expressed in [1]–[8], we've captured the heart of the narrativity thesis. So it hasn't told us anything we didn't always know or indeed anything that isn't obvious. Claims about narrativity are pleasing to many ears. They sound newly substantive, interestingly textured. It may be, though, that they sound good simply because they have all these obvious facts behind them, but don't really add anything to them. So they're like fool's gold.

One of the ironies in the debate is that it is precisely the most passionate defenders of the narrativist approach who tend to push it into triviality. Their determination to establish narrativity as a universal feature of non-pathological human life encourages them to treat features of human life whose universality no one can plausibly deny as if they were already sufficient for (constitutive of) narrativity, thereby undermining the potential interest and distinctiveness of the claim.[69] Some of them seem to think that the *Locke* point already secures the psychological narrativity thesis. But—again—if possessing the Locke property already entails being narrative, the claim that all normal people are narrative is empty, and we certainly don't need to use the word "narrative" to express it. All we need to do is to say, with Locke, that people can "consider [themselves] as [themselves], the same thinking thing, in different times and places."

If on the other hand the *Locke* property doesn't entail being narrative, the claim that all normal people are narrative is so far without support, and seems false, because it seems plain that some of us don't naturally cast or construe our lives as a narrative or story of some sort and don't experience parts of our lives in this way either. We don't think of our lives in this way even though the *Locke*, *Knowledge 101*, *Self-History*, *Timeline*, *Psychology*, *Explanation*, and *Action* points apply to us. Nor do we think of ourselves (as opposed to our lives) in a narrative way, whatever exactly this might be supposed to involve. We may simply not be reflective about these things, or we may be like Bob Dylan:

> I don't think I'm tangible to myself. I mean, I think one thing today and I think another thing tomorrow. I change during the course of a day. I wake and I'm one person, and when I go to sleep I know for certain I'm somebody else. I don't know who I am most of the time. It doesn't even matter to me.[70]

Narrativists may say that Dylan's "I'm one thing today, another thing tomorrow" is just part of his "narrative." This is another emptiness move, because it equates self-narrative with self-conception. In fact it's not clear that everyone has some sort of definite characteral self-conception. Sartre thinks it's plainly false. But even if it were true, simply equating self-narrative with self-conception would make the psychological narrativity thesis empty.

When we place the psychological narrativity thesis and narrative self-constitution thesis alongside the *Locke*, *Timeline*, and *Self-History* points, I think we can see part of what these points are meant to add. At the very least, they're meant to add the idea that

[a] one engages in a certain activity of "form-finding" directed at one's life,[71] a certain sort of structured selection from things one knows about one's life and oneself, a certain sort of shaping of these elements into *a temporally ordered and connected picture* of who one is and what one's life amounts to, and

[b] one's sense of oneself and one's life is deeply structured by this activity, which is in that respect fundamentally constitutive of who one is.

What about this?

The first thing to say, perhaps, is that if [a] is to have any empirical plausibility as a thesis about the generality of humankind, it will have to be possible for this form-finding to be largely and sometimes wholly unconscious.

Suppose we allow the unconsciousness, if only for the sake of argument, and take the notion of unconsciousness in a literal and theoretically neutral way: an activity like form-finding is unconscious if and only if it's not conscious. Suppose we allow in addition that such form-finding is universal, and that everyone has some sort of fairly determinate self-conception, and that it is significantly shaped by some sort of form-finding. We still face a very large doubt about whether there is anything distinctively *narrative* about the form-finding and equally the self-conception. If certain things about one's life stand out for one, one isn't ipso facto narrative. There's nothing specifically narrative about this process (cf. *Locke*, *Self-History*, *Timeline*). The things that stand out are highly likely to be isolated episodes, and their standing out as they do will be a function of how one already is.

Dramatic experiences may change a person in various ways, but there isn't anything specifically narrative about this process. And the fundamental psychological constitution of a human being doesn't change significantly over time, short of trauma or radical neural disorder. Really there's little or nothing one can do about one's fundamental psychological constitution other than hope to make progress toward its best expression. And this, perhaps, is largely a matter of luck.

Let me now take back one of the suppositions I just made for the sake of argument. People's self-conceptions are usually fantastically muddy, out in the real world—and don't generally seem to be narrative phenomena in any interesting sense. When David Brooks says "We work hard to cram our lives into legible narratives. But we live in the fog of reality,"[72] the first sentence describes nothing I know. The second does. I think some of us simply live in the fog of reality (or, less pejoratively, the thick texture of reality) and never do any narrative cramming. Like everyone else, I sometimes think back to past events and wonder what exactly happened. But I can't be said to be narrative (nonemptily) simply because I think back and wonder whether what I said to Tallulah was the best thing to say in the circumstances. Perhaps the only deeply narrative types are the types who write about narrativity and argue falsely that we're all narrative. In his autobiographical book *Coda*, the psychologically subtle playwright Simon Gray realizes that "the truth is that I don't really know even quite elementary things about myself, my wants and needs, until I've written them down or spoken them."[73] There may be, for all that, a sense in which he knows himself well, and better than many urgent self-storyers.

What else might the psychological narrativity thesis add to *Locke*, *Timeline*, and *Self-History*? Perhaps the idea of fiction: many agree with Ricoeur that narrativity essentially involves a certain sort of self-fictionalizing activity. Again, this seems unfortunate if true, given that it's usually better to stick to the facts. But this doesn't matter here. What matters is simply that it doesn't seem to be true that everyone does this.

"It is true," the narrativist objects. "All ordinary people are self-deceived in various ways, and self-deception is an *essentially* narrative activity. So the psychological narrativity thesis is true."

I agree that self-deception is almost universal, even perhaps universal, but now the threat of emptiness returns. We're essentially temporal creatures, so self-deception necessarily concerns things that are situated in time. But it can't be said to be *narrative* (in any nonempty sense) simply because it concerns matters that are situated in time. And if the term "narrative" is introduced simply because it brings with it the idea of fiction and hence (in this case) falsification, its use is question-begging. For who is to say that self-deception can't be false, fiction, without being specifically narrative in character?

We're all affected and shaped by our past, although our basic natures are set early on. We need to ask how much of this change is mediated by activity that qualifies as narrative activity from the narrativists' point of view. And this raises another question: how much of this change is supposed to be something we do, rather than something that just happens, like a scar after a wound? And now this question arises: if some of it just happens, how much of it happens in a way that has a distinctively narrative character, and how much in a way that has no distinctively narrative character? And this raises another question: if some of it (or all of it) is something we do, how much of it is something we consciously do and how much something we unconsciously do? And this in turn raises a doubt: how plausible is this notion of unconscious doing as opposed to unconscious happening? And now another question: how much of this unconscious activity, whether it is thought of as doing or happening, is rightly thought of as a distinctively narrative activity?

Suppose narrativists allow that it's mostly unconscious—unconscious in the plain sense of nonconscious. Suppose they agree that Schechtman is right to retreat from the view that self-constituting self-narration is paradigmatically an explicit conscious affair to the view that it is largely implicit and nonconscious.[74] Suppose we in return allow (while expressing serious reservations) that the unconscious goings-on may be in some sense narrative in character, even if they're something that happens rather than something we do. Now we must wonder about the involvement of the unconscious in the nonplain sense: not only the "Freudian unconscious" (the uppercase Unconscious, the "System Ucs") but also the non-Freudian "adaptive unconscious" charted by Wilson, Kahneman, and others.[75] Why should we think that the agendas and effects of the Freudian unconscious and the adaptive unconscious count as narrative?

I think this question goes far and deep. It's not just a matter of questions like "Is one essentially narrative in nature because something that happened to one when one was three still affects how one feels about fur coats?"—to which the answer must be "No." For there is also, and again, nothing distinctively *narrative* about an experience of sudden insight into oneself—"I realize that I felt at peace at that moment in a way I have never done before"; "I can see now that what I really wanted was *x*"; "What I did then reveals the truth about what I'm really like"—whether the experience is accurate or not.

A Methodological Proposal

I suspect that more than 95 percent of all supposedly descriptively substantive theoretical uses of the noun "narrative" ("story," "tale") in ethics and psychology can be replaced without any loss by words like "explanation," "description," "account," "view," "outlook," "theory," "understanding," "theme," "belief," "concept," "conception," and "picture."[76] Let me make a methodological proposal: whenever anyone uses the word "narrative" or "story," ask whether it can be replaced without semantic loss by one of the words just listed or by some variant of the generic expression "description or account of a temporally extended sequence of events." If it can be replaced, let it be replaced. Then we'll be able to see more clearly what's left. If there's disagreement about whether something has been lost, the burden of proof will fall on the user of the word "narrative."

Take Timothy Wilson's excellent book *Strangers to Ourselves* (2002). In addition to talking regularly of "narrative" and "self-narrative,"[77] Wilson also talks of a person's "self-concept,"[78] "self-theory,"[79] and "self-view."[80] Let's assume again that every normal person has some sort of reasonably substantive self-concept, self-theory, or self-view. The present point is then that "self-concept," "self-theory," and "self-view" can't be taken to be effectively synonymous with "self-narrative." For then again the narrativity thesis risks emptiness. Someone who wishes to defend the narrativity thesis must *argue* that any substantive self-concept/theory/view is specifically narrative in character, at least in some fundamental respect.[81] They must also show that it is narrative in some sense that essentially transcends the eight points listed earlier—otherwise their claim will again be empty. They will also have to argue for the contentious claim that every normal human being has a substantive self-concept/theory/view.[82]

At one point Wilson quotes Robert Zajonc, "an eminent social psychologist who has never quite understood all the fuss about self-psychology. 'I don't think of my self, as such,' he once said, 'I may think of my schedule, my obligations, my meetings, but I don't really spend too much time asking 'Who am I?'"[83]

It's important to put such cases on record. But Wilson continues as follows: "He [Zajonc] may be on to something. A good self-narrative need not be constantly retold." Here Wilson simply assumes that a self-view is a self-narrative, and so begs the question raised by this chapter.

Zajonc is not alone. We've heard Bob Dylan. Here is the philosopher Freddie Ayer:

> I wonder whether I do even have an image of myself? I suppose if you ask me questions about my character, I would have opinions about most of them, I would say: "Yes, I am X," or "I'm not Y" [but] I don't think that I have an image of myself in the sense that I am much concerned with my own character. I don't think I'm all that much interested in myself. I'm much more interested in trying to solve philosophical problems, or in what's happening around me, or how my little boy is getting on.[84]

If even Ayer is inevitably and essentially narrative, the term has no teeth—it's empty. Some people may lack a clear self-concept because of a mental disorder, but the lack is not itself a mental disorder.

Empirical Investigation

The point extends to Wilson's use of the term "self-biography."[85] I think it's plainly false that all ordinary human beings' self-biographies are informed by a distinctively storytelling or narrative bent. We should consider the matter empirically (i.e., each of us should consider himself or herself). A fallible method of investigation, no doubt, but there's no reason to think that it isn't good enough for this purpose, given that we haven't shunted narrativity into the unconscious.[86]

If I consider myself, I find that my self-biography is just a chronology, a list of dates. It's a filing cabinet (mostly empty). It doesn't in any way represent progression. It involves no narrative flow, although I can of course report certain causal sequences (but I am bad at this). It does mark a few "moments of being," in Virginia Woolf's sense, events that I know occurred in my life and about which I feel strong emotions. But I have almost no clear memory of what happened and in that sense remember them only in a highly abstract way. These are, precisely, *moments* of being, snapshots, fundamentally nonnarrative within themselves,[87] and not phenomenologically part of any larger containing sequence or "narrative."

What about my self-concept, my self-conception? When I try to summon it, I have a sense of complete blankness. Insofar as I come up with something, it seems again perfectly nonnarrative, profoundly nontemporal, a kind of shape, a feeling of a form. There's no possible place in it for anything as busy as narrative, anything discursive or temporally segmented or extended. My lack or loss, you may say. My difference, I reply—but I think it may be like this for many people. One doesn't have to go as far as Proust and his narrator in *A la recherche du temps perdu* and think that one's essence is outside time.

My reaction, then, is one of bewilderment. I don't think I'm revealing some moral or spiritual inferiority in admitting this. Nor am I covertly claiming moral or spiritual superiority. I'm recording a human data point of the sort that needs to be recorded as the levels of narrababble rise around us. My sense is that all that I am is here now and that what I am is fundamentally unclear to me. It's a profoundly nonnarrative experience.

If I try to think further about myself—if I try to bring a self-conception to mind—initial thoughts about character traits are met with blankness, confirming Sartre's observation that "character has no distinct existence except as an object of knowledge to other people. Consciousness does not know its own character—except in so far as it may consider itself reflectively from the point of view of another.... Pure introspective description of oneself does not reveal a character."[88] I start thinking about how I feel about and respond to others: how I react to items on the news, and the explosion, in the last ten years, in the pleasure I take in wild flowers. I encounter items that seem tangential but may be informative for that reason: I enjoy modest risk-taking; I like taking baths in the dark; I don't like exercise; I like feeble jokes and try to make them.

Such facts—bits and pieces—may be important. Some of them may be profoundly diagnostic of identity, Proustian keys, keys to identity, to Flanagan's AFI. To take them for what they are—bits and pieces—and not to try to assemble them in any discursive fashion, may be the beginning of wisdom, and perhaps also the end of it, in any project of self-understanding. Together they may suggest an individual shape, a shape that one may have a chance of discerning to some degree, in spite of Nietzsche's skepticism, and especially if one doesn't try to see it more clearly.[89]

Now I think about scheduling appointments with students; I think about my reactions to bad drivers; I wonder if I have done enough for a certain person. I think my capacity to hate is weak; this triggers a reflex La Rochefoucauldian suspicion about my knowledge of my deep dispositions: what is my capacity for self-deception?[90] This is standard stuff: attempts at self-assessment routinely push us into the moral mode.

If I go on, thoughts about how I've behaved in the past may come in to test my judgments about my character. But they're not a matter of narrative—unless considering events in one's past in this way is stipulated to be a matter of narrativity. Any such stipulation would again render the psychological narrativity thesis empty in the current sense, insofar as it's true at all (insofar as it's true that all normal human beings consider events in their past in such cases).

Objection: What comes up for you, what is available for memory, is a matter of all the narrative work you have already unconsciously done.

Reply: Many unconscious processes have contributed to what I can and cannot remember. I don't know how "motivated" these processes are, if only because my personal memory is extraordinarily poor across the board (trouble in the hippocampus?). But suppose the forgettings are "motivated" in the sense of being driven by a certain preferred self-conception. Why say that this largely inaccessible self-conception is a self-*narrative*? Perhaps you want to call it a self-narrative because you think any such self-conception is partly *constituted* by a certain set of personal memories, a set that may itself be said to constitute a narrative simply insofar as the remembered events have a certain temporal order (which I may be wrong about) and centrally concern a single person. On this view, everyone is narrative, given that everyone has some sort of self-conception, and the psychological narrativity thesis is once again trivially true, and therefore empty in the present sense, for now all self-conceptions are self-narratives, and we have assumed that everyone has some sort of self-conception more or less substantive, more or less immediately available to consciousness.

I don't think that a self-conception is well thought of as partly constituted by a set of memories. And even if we were to countenance this idea, it would be wholly unjustified to say that a set of memories constitutes a narrative simply because the remembered events have a certain temporal order and concern a single person. We live, in any case, in V. S. Pritchett's words, "beyond any tale that we happen to enact,"[91] and any truly worthwhile self-conception will have to be true to that fact. It follows that it will need to have a radically nonnarrative form.

Conclusion

This is the beginning of an attempt to ask one of the core questions in this debate. What is one for oneself? What is one's sense of oneself? The specific question is: must this fundamental sense of oneself have a distinctively narrative character? As far as almost all conscious goings-on are concerned, the answer is certainly no. Does it ever have this character? I sometimes doubt it, although I hope I never underestimate human difference. John Updike seems right when he says that "our own minds ... know ourselves," to the limited extent that they do, "as an entity too vast and vague to name."[92] Information about one's past life—one's history—may be more or less ready to hand, often murkily, as a source of data for testing one's views about oneself. It may be somewhere in one's personal internal "cloud." But—for many at least, if not most—it's not part of what one is for oneself. What matters most to a great many of us, I think, is a certain overall sense of our moral identity as we are now in the present moment—realistic or otherwise.[93]

Obviously, my history has played a deep part in making me what I am, where what I am includes my general sense of myself. But it's no more part of my sense of myself than my past visual experiences of my childhood bedroom are part of my present visual experience of the room I'm in now. So it seems to me. They're part of my history, and they're part of the causal history that lies behind my present being, but they're not part of my narrative, where this is a narrative I tell myself about myself, because I don't have any such narrative. I am not (in Bellow's terms) exiled from myself in this way.

It's principally in the eyes of others that one's life or history is part of what one is—far more so than it is for oneself. This is how it is for me, at least, and I think I'm one of many. Plainly, we need more research on individual differences. We need to stand up to all these wildly sweeping generalizations about the universality and necessity of narrativity. Perhaps one of the things that is going wrong in this debate is that some narrativists, swept away by theory, have fallen into an excessively public picture of a human being's self-conception. Perhaps they've forgotten what life is actually like—the inner life, in particular, which is so vast a part of any ordinary human life (especially for those who lie awake at night). Perhaps the narrativists, dazed by theory, are modeling self-experience falsely on the third-person point of view, the view from the outside. Or perhaps this is what it's like for them (it seems a kind of self-alienation, as Sartre says). In that case, they have overgeneralized from their own cases. If they're telling the truth about the character of their self-experience, then they're astonishingly different from me.[94]

Notes

1. Greenblatt, Shakespeare's Montaigne, xiv.

2. Montaigne, *The Complete Essays*, 120.

3. For some recent interesting work, see Palombo et al., Severely deficient autobiographical memory (SDAM) in healthy adults; Watkins, (A)phantasia and severely deficient autobiographical memory. I'm principally concerned with differences in "personal," "episodic," or autobiographical memory, but there's no need to exclude other forms of memory. Klein argues for a restriction of the theoretical use of the term "memory" to the case of personal memory; see Klein, What memory is.

4. Owen Flanagan stresses the importance of taking proper account of variation; see Flanagan, *Varieties of Moral Personality*.

5. Wiesel, *All Rivers Run to the Sea*, 150.

6. Frisch, *What Little I Remember*, ix.

7. Bellow, *The Bellarosa Connection*, 2. He allows that "in the universal process of self-revision anybody can be carried away from the true facts" (ibid.).

8. Montaigne, *The Complete Essays*, 32; "Memory is an instrument of wondrous service.... In me it is entirely lacking" (ibid., 796). I don't know whether Montaigne and I qualify as cases of SDAM, severely deficient autobiographical memory.

9. Ryle, *The Concept of Mind*, 279.

10. Halpern, *Can't Remember What I Forgot*, 66.

11. Hacking, discussing Ryle, holds "that one becomes good at recalling the critical part of one's past when one acquires the skill to cast it into a coherent narrative." See Hacking, *Rewriting the Soul*, 256. Again, I think this is the opposite of the truth.

12. Halpern, *Can't Remember What I Forgot*, 21.

13. "For the sake of a single poem ... , you must have ... many ... memories.... And yet it is not enough to have memories.... For the memories themselves are not important." They give rise to a good poem "only when they have changed into our very blood, into glance and gesture, and are nameless, no longer to be distinguished from ourselves." See Rilke, *The Notebooks of Malte Laurids Brigge*, 91.

14. Simic, Grass, 23.

15. Sacks, *The Man Who Mistook His Wife for a Hat*, 110.

16. "A self is probably the most impressive work of art we ever produce, surely the most intricate. For we create not just one self-making story but many of them, rather like T. S. Eliot's rhyme 'We prepare a face to meet / the faces that we meet.'" See Bruner, The "remembered" self, 53. "The job," according to Bruner, "is to get them all into one identity, and to get them lined up over time" (Bruner, *Making Stories*, 14). I know of no such job.

17. MacIntyre, *After Virtue*, 197 (3rd ed. 212).

18. Taylor, *Sources of the Self*, 52.

19. Hardy, Towards a poetics of fiction, 5.

20. Schechtman, *The Constitution of Selves*, 93. She has since considerably modified her view.

21. McAdams, Josselson, and Lieblich, Introduction, 3.

22. Velleman, The self as narrator.

23. Flanagan, *Consciousness Reconsidered*, 192.

24. Flanagan, Multiple identity, character transformation, and self-reclamation, 139.

25. Pullman, personal communication.

26. Novalis, *Novalis*, 66.

27. "Un homme, c'est toujours un conteur d'histoires, il vit entouré de ses histoires et des histoires d'autrui, il voit tout ce qui lui arrive à travers elles; et il cherche à vivre sa vie comme s'il la racontait. Mais il faut choisir: vivre ou raconter" (Sartre, *La nausée*, 64).

28. Sartre, *La nausée*, 64.

29. Barnes, *Nothing to Be Frightened Of*, 152.

30. Murdoch, Metaphysics and ethics, 75.

31. Murdoch, The sovereignty of good over other concepts, 93.

32. McAdams, Josselson, and Lieblich, *Identity and Story*, 3.

33. Sacks, *The Man Who Mistook His Wife for a Hat*, 110.

34. Bruner, Life as narrative, 15.

35. Hardy, Towards a poetics of fiction, 6. The doyen of this approach is social psychologist Dan McAdams. He gives us what he calls the "*life-story theory of identity*" (McAdams, The Redemptive Self, 95). "*Identity*," he says, " *is itself a life story*" (ibid., 99). "Beginning in late adolescence and young adulthood we construct integrative narratives of the self that selectively recall the past and wishfully anticipate the future to provide our lives with some semblance of unity, purpose, and identity. Personal identity is the internalized and evolving life story that each of us is working on as we move through our adult lives.... I ... do not really know who I am until I have a good understanding of my narrative identity" (McAdams, *The Redemptive Self*, 287–288). Marya Schechtman states the strong view in an exemplary form: "A person's identity is created by a self-conception that is narrative in form.... This means that constituting an identity requires that an individual conceive of his life as having the form and the logic of a story—more specifically, the story of a person's life—where 'story' is understood as a conventional, linear narrative.... We constitute ourselves as persons by forming a narrative self-conception according to which we experience and organize our lives" (Schechtman, Stories, lives, and basic survival, 162). In her most recent work, however, she weakens these claims considerably, and concludes by saying that "it seems more accurate and less liable to generate misunderstanding to give up the locution of 'narrative' in this context and to describe the type of unity that defines a person's identity not as a *narrative* unity but simply as the structural unity of a person's life" (Schechtman, *Staying Alive*, 108).

36. McAdams, First we invented stories, then they changed us.

37. Flanagan, Multiple identity, character transformation, and self-reclamation, 139–140.

38. Flanagan, My non-narrative, non-forensic *dasein*, 220. Flanagan has recently defined AFI somewhat differently, as "the true and complete story of who a person was, how he lived, what he did, and why" (Flanagan and Graham, Truth and sanity, 112). Peter van Inwagen once told me that he thinks that this "story" (in fact a wholly true and wholly factual description) may be what one learns on the Day of Judgment, when "the secrets of all hearts shall be laid open" (1 Cor. 14:25), quoted in Locke, *An Essay Concerning Human Understanding*, 2.27.26, and that one may then have to do something that might be almost impossibly difficult: accept that it is correct.

39. The notion remains robust in the face of all the spectacular "situationist" facts that are sometimes thought to undermine it. See, e.g., Doris, *Lack of Character*; Zimbardo and Sommers, The power of the situation.

40. Murdoch, The idea of perfection, 1.

41. I don't think this is Socrates' view, in fact. It's not at all clear that one has to think much *about oneself* in order to live the examined life.

42. Ricoeur, *Oneself as Another*, 114.

43. Didion, On keeping a notebook, 134.

44. Didion, *The White Album*, 1; Amis, Joan Didion's style, 4.

45. Taylor and Brown, Positive illusion and well-being. For doubts, see Van Leeuwen, Self-deception won't make you happy. See also Flanagan and Graham, Truth and sanity, 292–313.

46. Descartes, Letter to Princess Elisabeth of Bohemia, 3.268.

47. Mario Vargas Llosa is one of many who stress this point (Vargas Llosa, In praise of reading and fiction).

48. I'm grateful to Michael Smith for making this point at a workshop in Melbourne.

49. It can be taken to be a representation of something, but that is true of absolutely everything.

50. Blattner, Life is not literature, 187.

51. Locke, *An Essay Concerning Human Understanding*, 2.27.9.

52. In fact one can be fully self-conscious even if one has an extremely short memory span and very little conception of the future. As it stands, Locke's definition implies that a person must have a more considerable temporal range.

53. Rudd, *Self, Value and Narrative*.

54. Some "people ... come to believe their own stories by constantly repeating them" (Leibniz, *New Essays on Human Understanding*, 2.27.16). "Liars ... by the frequent repetition of their lies, come at last to believe and remember them, as realities" (Hume, *A Treatise of Human Nature*, 1.3.9.19).

55. It gives great force to Bellow's comment on Mr. Sammler: "He didn't in fact appear to know his age, or at what point in life he stood. You could see that in his way of walking." See Bellow, *Mr. Sammler's Planet*, 6.

56. "Our thoughts and enquiries are ... every moment, employed about [the] relation ... of cause and effect" (Hume, *Enquiries Concerning Human Understanding*, sec.7.29).

57. Why not "invariably"? A cannonball sitting on a springy cushion is the cause of the dent in the cushion.

58. Kant, *Critique of Practical Reason,* B288.

59. People vary enormously in how much planning they do, in the scope of their plans, and also, I suspect, in how they experience planning. For some people it's oddly impersonal or abstract. The fact that it concerns themselves in the future is given in the planning, but there's no vivid representation of this fact of a sort that might support the idea that planning involves a distinctively narrative outlook on their own lives. For others, perhaps, planning is vivid in precisely this way.

60. My emphasis. See MacIntyre, *God, Philosophy, Universities,* 396.

61. Schechtman, The narrative self, 396.

62. Rudd, *Self, Value and Narrative,* 197.

63. Ibid., 183. The second emphasis is mine.

64. Compare with Rudd, Summary of *Self, Value and Narrative*: "Insofar as selves are temporal and are agents (i.e., beings that act, and do so for reasons) they must understand themselves in narrative terms." As Zahavi says, "the term ['narrative'] threatens to become all-inclusive and consequently vacuous—in the end everything meaningful involves narratives—and this is surely a sign of bankruptcy" (Zahavi, Self and other, 196).

65. Rudd, *Self, Value and Narrative,* 197.

66. There is an extended descent into emptiness in the current sense in Rudd, *Self, Value and Narrative,* 177.

67. Ibid., 186.

68. Some say that every explanation of a human action is a narrative. Some say that every explanation of anything that happens to a human being is a narrative. Some say that every explanation of anything is a narrative. Some say that any intention to do anything involves a narrative. The theoretical trash (the emptiness) piles up.

69. Flanagan also notes this; see Flanagan, My non-narrative, non-forensic *dasein,* 236.

70. Dylan interview.

71. Strawson, Against narrativity.

72. Brooks, The courage of small things.

73. Gray, *Coda,* 114. Gray began the book after learning that he was dying of cancer.

74. See, e.g., Schechtman, *Staying Alive*, 100–101.

75. See, e.g., Wilson, *Strangers to Ourselves*; Wilson, Knowing when to ask; Wilson, *Redirect*; Kahneman, *Thinking, Fast and Slow*. For many anticipations of this work, see also Bacon, *Novum Organum Scientiarum*; Hume, An *Enquiry Concerning Human Understanding*.

76. This won't include uses of "narrative" and its cognates that simply name the position, like "the narrative approach" or "narrative theory." The adjective "narrative" can quite often be replaced by "psychological" or "reasons-citing."

77. See, e.g., Wilson, *Strangers to Ourselves*, 90, 159–181, 210, 213, 216–221.

78. Ibid., 69, 86–87, 195, 214.

79. Ibid., 195–200.

80. Ibid., 38, 52, 86, 196–199, 215, 218, 221.

81. A narrativist can perhaps allow that someone can have a non-narrative self-view in addition to having a self-narrative.

82. It seems to me that every occurrence of "narrative" in Wilson's equally excellent book *Redirect* is smoothly eliminable. The same goes for occurrences of the term "story-editing," although it is particularly apt in chapter 1, which shows how one can change students' performance by changing their expectations of how they will do. John Davenport proposes the following "Signature Narrative Thesis: each person's individual identity *is*, or *depends on*, an understanding he has of his life in narrative form, as a development from his past towards his future prospects, ending in his death" (Davenport, *Narrative Identity and Autonomy*, 2). One can delete the words "in narrative form" without losing anything. The claim remains false. The italicized "is" option seems barely intelligible, and obviously false insofar as it is intelligible. The italicized "depends on" option seems intelligible but empirically false, unless we allow that it's made true simply by the fact that normal human beings know the normal human life span together with the fact that this knowledge conditions their outlook on life in some way. In this case, the claim is again trivially true and so empty in the present sense.

83. Wilson, *Strangers to Ourselves*, 221.

84. Quoted in Rogers, *A. J. Ayer*, 286. See also Bayley, *Iris*, 51–52: "Iris [Murdoch] once told me that the question of identity had always puzzled her. She thought she herself hardly possessed such a thing, whatever it was. I said that she must know what it was like to be oneself, even to revel in the consciousness of oneself, as a secret and separate person—a person unknown to any other. She smiled, looked amused, uncomprehending. It was not something she bothered about."

85. Wilson, *Strangers to Ourselves*, 163, 183, 219–220.

86. Shunting narrativity into the unconscious may make the psychological narrativity thesis effectively indefeasible—but also indefensible, given the difficulty of establishing that the processes of the unconscious are distinctively narrative in character.

87. They may be "clips" (or "gifs," as Engel Hawbecker has suggested) rather than snapshots and contain representation of movement, like Virginia Woolf's childhood memory of the "hearing the waves breaking, one, two, one, two, ... hearing the blind draw its little acorn across the floor as the wind blew the blind out" (Woolf, A sketch of the past, 64–65). But that had better not be enough for them to qualify as narrative, on pain of emptiness.

88. Sartre, *Being and Nothingness*, 349.

89. "How can man know himself? He is a dark and veiled thing; and whereas the hare has seven skins, the human being can shed seven times seventy skins and still not be able to say: 'This is really you, this is no longer outer shell'" (Nietzsche, Schopenhauer as educator, 174).

90. This in turn triggers speculation about how I'd judge someone else who'd behaved as I have; many people judge themselves much more harshly than they judge others. This self-harshness may well be a moral failing, a kind of self-indulgence, and also, perhaps more importantly, a failure of realism.

91. Pritchett, *The Myth Makers*, 47.

92. Updike, *Gertrude and Claudius*, 76.

93. "The metaphysicians ... affirm that if memory be taken away, the self is lost. And what matter for memory? What have I to do with that part? If, whilst I am, I am as I should be, what do I care more? And thus let me lose self every hour, and be twenty successive selfs, or new selfs, "tis all one to me: so I lose not my opinion [i.e., my overall outlook, my character, my moral identity]. If I carry that with me "tis I; all is well....—The now; the now. Mind this: in this is all" (Shaftesbury, Philosophical Regimen 136–137).

94. Special thanks to Michael Smith, and to Stan Klein, Tom Polger, and Eddy Nahmias for their comments.

Bibliography

Amis, Martin. Joan Didion's style. *London Review of Books* 2, no. 2 (February 1980): 3–4.

Bacon, Francis. *Novum Organum Scientiarum* (1620). Translated by L. Jardine and M. Silverthorne. Cambridge: Cambridge University Press, 2000

Barnes, Julian. *Nothing to Be Frightened Of*. London: Jonathan Cape, 2008.

Bayley, John. *Iris: A Memoir*. London: Duckworth, 1998.

Bellow, Saul. *The Actual*. New York: Viking, 1997.

Bellow, Saul. *The Bellarosa Connection*. London: Penguin, 1989.

Bellow, Saul. *Mr. Sammler's Planet*. New York: Viking, 1970.

Blattner, William. Life is not literature. In *The Many Faces of Time*, edited by L. Embree and J. Brough, 187–201. Dordrecht: Kluwer, 2000.

Brooks, David. The courage of small things. *New York Times*, July 7, 2015, A19.

Bruner, Jerome. Life as narrative. *Social Research* 54, no. 1 (1987): 11–32.

Bruner, Jerome. *Making Stories: Law, Literature, Life*. Cambridge, MA: Harvard University Press, 2003.

Bruner, Jerome. The "remembered" self. In *The Remembering Self: Construction and Accuracy in the Self-Narrative*, edited by U. Neisser and R. Fivush, 41–54. Cambridge: Cambridge University Press, 1994.

Davenport, John. *Narrative Identity and Autonomy: From Frankfurt and MacIntyre to Kierkegaard*. London: Routledge, 2012.

Descartes, René. Letter to Princess Elisabeth of Bohemia, October 6, 1645. In *The Philosophical Writings of Descartes*, vol. 3, translated by John Cottingham, Robert Stoothoff, and Dugald Murdoch. Cambridge: Cambridge University Press, 1991.

Didion, Joan. On keeping a notebook (1966). In *Slouching toward Bethlehem*, 131–141. New York: Farrar, Straus and Giroux, 1968.

Didion, Joan. *The White Album*. New York: Farrar, Straus and Giroux, 1979.

Doris, John. *Lack of Character: Personality and Moral Behavior*. New York: Cambridge University Press, 2002.

Dylan, Bob. Interview. *Time*, October 6, 1997.

Flanagan, Owen. *Consciousness Reconsidered*. Cambridge, MA: MIT Press, 1992.

Flanagan, Owen. Multiple identity, character transformation, and self-reclamation. In *Philosophical Psychopathology*, edited by George Graham and Lynn Stephens, 135–162. Cambridge, MA: MIT Press, 1994.

Flanagan, Owen. My non-narrative, non-forensic *dasein*: The first and second self. In *Consciousness and the Self: New Essays*, edited by Jeeloo Liu and John Perry, 214–240. Cambridge: Cambridge University Press, 2011.

Flanagan, Owen. *Varieties of Moral Personality: Ethics and Psychological Realism*. Cambridge, MA: Harvard University Press, 1991.

Flanagan, Owen, and Graham, George. Truth and sanity: Positive illusions, spiritual delusions, and metaphysical hallucinations. In *Extraordinary Science and Psychiatry: Responses to the Crisis in Mental Health Research*, edited by Jeffrey Poland and Şerife Tekin, 293–314. Cambridge, MA: MIT Press, 2017.

Frisch, Oskar. *What Little I Remember*. Cambridge: Cambridge University Press, 1979.

Gray, Simon. *Coda*. London: Faber and Faber, 2008.

Greenblatt, Stephen. Shakespeare's Montaigne. In *Shakespeare's Montaigne: The Florio Translation of the Essays*, edited with an introduction by S. Greenblatt and edited, annotated, and modernized by P. Platt, ix–xxxiii. New York: New York Review of Books, 2014.

Hacking, Ian. *Rewriting the Soul: Multiple Personality and the Sciences of Memory*. Princeton, NJ: Princeton University Press, 1998.

Halpern, Sue. *Can't Remember What I Forgot: Your Memory, Your Mind, Your Future*. 2008. Reprint, New York: Three Rivers Press, 2009.

Hardy, Barbara. Towards a poetics of fiction: An approach through narrative. *Novel* 2 (1968): 5–14.

Hume, David. An *Enquiry Concerning Human Understanding* (1748–1751). Edited by L. A. Selby-Bigge and P. H. Nidditch. Oxford: Oxford University Press, 1975.

Hume, David. *A Treatise of Human Nature* (1739–1740). Edited by D. F. Norton and M. Norton. Oxford: Clarendon Press, 2000.

Kahneman, Daniel. *Thinking, Fast and Slow*. New York: Farrar, Strauss, Giroux, 2011.

Kant, Immanuel. *Critique of Pure Reason* (1781–1787). Translated by N. Kemp Smith. London: Macmillan, 1933.

Klein, Stanley. What memory is. *Wiley Interdisciplinary Reviews Cognitive Science* 6, no. 1 (August 2015): 1–38. https://doi.org/10.1002/wcs.1333.

Leibniz, Gottfried. *New Essays on Human Understanding* (c. 1704), edited and translated by J. Bennett and P. Remnant. Cambridge: Cambridge University Press, 1996.

Locke, John. *An Essay Concerning Human Understanding*. 2nd ed. London: Dring and Manship, 1694.

MacIntyre, Alasdair. *After Virtue* (1981). 3rd ed., Notre Dame, IN: University of Notre Dame Press, 2007.

MacIntyre, Alasdair. *God, Philosophy, Universities: A Selective History of the Catholic Philosophical Tradition*. Lanham, MD: Rowman and Littlefield, 2011.

McAdams, Dan. First we invented stories, then they changed us: the evolution of narrative identity. *Evolutionary Studies in Imaginative Culture* 3, no. 1 (2019): 1–18.

McAdams, Dan. The Redemptive Self: Narrative Identity in America Today. In *The Self and Memory*, edited by D. R. Beike, J. M. Lampinen, and D. A. Behrend, 95–115. New York: Psychology Press, 2004.

McAdams, Dan, Josselson, Ruth, and Lieblich, Amia. Introduction. In *Identity and Story: Creating Self in Narrative*, 1–11. Washington, DC: American Psychological Association, 2006.

Montaigne, Michel de. *The Complete Essays (*1563–1592*)*. Translated by M. A. Screech. London: Penguin, 1991.

Murdoch, Iris. The idea of perfection (1964). In *The Sovereignty of Good*, 1–45. London: Routledge and Kegan Paul, 1970.

Murdoch, Iris. Metaphysics and ethics (1957). In *Existentialists and Mystics*, 59–75. London: Penguin, 1997.

Murdoch, Iris. The sovereignty of good over other concepts (1967). In *The Sovereignty of Good*, 77–104. London: Routledge and Kegan Paul, 1970.

Nietzsche, Friedrich. Schopenhauer as educator (1874). In *Unfashionable Observations*, translated by R. Gray. Stanford, CA: Stanford University Press, 1997.

Novalis. Logological Fragments I. In *Novalis: Philosophical Writings* (c. 1798), edited and translated by M. Stoljar, 47–66. Albany: State University of New York Press, 1997.

Palombo, Daniela J., Alain, Claude, Söderlund, Hedvig, Khuu, Wayne, and Levine, Brian. Severely deficient autobiographical memory (SDAM) in healthy adults: A new mnemonic syndrome. *Neuropsychologia* 72 (June 2015): 105–118.

Pritchett, Victor. *The Myth Makers*. London: Chatto and Windus, 1979.

Ricoeur, Paul. *Oneself as Another*. Translated by Kathleen Blamey. 1990. Reprint, Chicago: University of Chicago Press, 1992.

Rilke, Rainer Maria. *The Notebooks of Malte Laurids Brigge* (1910). Translated by S. Mitchell. New York: Random House, 1983.

Rogers, Ben. *A. J. Ayer*. London: Chatto and Windus, 1999.

Rudd, Anthony. *Self, Value and Narrative: A Kierkegaardian Account*. Oxford: Oxford University Press, 2012.

Rudd, Anthony. Summary of *Self, Value and Narrative: A Kierkegaardian Account*. Academia, 2012. https://www.academia.edu/1302730/Self_Value_and_Narrative_a_Kierkegaardian_Approach.

Ryle, Gilbert. *The Concept of Mind*. Harmondsworth: Penguin, 1949.

Sacks, Oliver. *The Man Who Mistook His Wife for a Hat*. London: Duckworth, 1985.

Sartre, Jean-Paul. *Being and Nothingness: An Essay on Phenomenological Ontology*. Paris: Éditions Gallimard, 1943.

Sartre, Jean-Paul. *La nausée* (1938). Paris: Gallimard, 1996.

Schechtman, Marya. *The Constitution of Selves*. Ithaca, NY: Cornell University Press, 1996.

Schechtman, Marya. The narrative self. In *The Oxford Handbook of the Self*, edited by Shaun Gallagher, 394–418. Oxford: Oxford University Press, 2011.

Schechtman, Marya. *Staying Alive*. Oxford: Oxford University Press, 2014.

Schechtman, Marya. Stories, lives, and basic survival. In *Narrative and Understanding Persons*, edited by Daniel D. Hutto, 155–178. Cambridge: Cambridge University Press, 2007.

Shaftesbury, Anthony. Philosophical regimen (1698–1712). In *The Life, Unpublished Letters, and Philosophical Regimen of Anthony, Earl of Shaftesbury*, edited by B. Rand, 1–266. New York: Macmillan, 1900.

Simic, Charles. Grass: The gold and the garbage. *New York Review of Books*, March 24, 2011.

Strawson, Galen. Against narrativity (2004). In *Real Materialism and Other Essays*, 189–207. Oxford: Oxford University Press, 2008.

Taylor, Charles. *Sources of the Self*. Cambridge: Cambridge University Press, 1989.

Taylor, Shelley, and Brown, Jonathon. Positive illusion and well-being: A social psychological perspective on mental health. *Psychological Bulletin* 103 (1988): 193–210.

Updike, John. *Gertrude and Claudius*. New York: Knopf, 2000.

Van Leeuwen, Neil. Self-deception won't make you happy. *Social Theory and Practice* 35, no. 1 (2009): 107–132.

Vargas Llosa, Mario. In praise of reading and fiction. Nobel Lecture, December 7, 2010. https://www.nobelprize.org/uploads/2018/06/vargas_llosa-lecture_en-1.pdf.

Velleman, J. David. The self as narrator. In *Self to Self: Selected Essays*, 203–223. Cambridge: Cambridge University Press, 2005.

Watkins, Nick. (A)phantasia and severely deficient autobiographical memory: Scientific and personal perspectives. *Cortex* 105 (2018): 41–52.

Wiesel, Elie. *All Rivers Run to the Sea*. New York: Knopf, 1995.

Wilson, Timothy. Knowing when to ask: Introspection and the adaptive unconscious. In *Trusting the Subject? The Use of Introspective Evidence in Cognitive Science*, edited by A. Jack and A. Roepstorff, 131–140. Thorverton, UK: Imprint Academic, 2003.

Wilson, Timothy. *Redirect: Changing the Stories We Live By*. London: Penguin, 2013.

Wilson, Timothy. *Strangers to Ourselves: Discovering the Adaptive Unconscious*. Cambridge, MA: Harvard University Press, 2002.

Woolf, Virginia. A sketch of the past (1939). In *Virginia Woolf: Moments of Being*, edited with an introduction by Jeanne Schulkind. London: Hogarth Press, 1978.

Woolf, Virginia. *The Waves*. London: Hogarth Press, 1931.

Zahavi, Dan. Self and other: The limits of narrative understanding. In *Narrative and Understanding Persons*, edited by Daniel D. Hutto, 179–202. Cambridge: Cambridge University Press, 2007.

Zimbardo, Philip, and Sommers, Tamler. The power of the situation. In *A Very Bad Wizard: Morality behind the Curtain*, edited by T. Sommers. San Francisco: Believer Books/McSweeney's Books, 2009.

6 Virtuous Women

Peggy DesAutels and Jack Bauer

> The right unit of attention for ethics is the whole person-in-communal relations, not person parts, say genes, or the emotional centers of the brain, or the rational parts of brains, not brains period, but persons who seek to live well in relations with other persons in particular natural and social ecologies with histories.
>
> —Owen Flanagan, *The Geography of Morals*

In an ideal world, women and men live virtuously in, and thrive while contributing to, flourishing communities. In our less-than-ideal world, however, persons' individual psychologies, virtues, and vices develop within and are facilitated and constrained by nonideal social conditions. In this chapter we focus on whether it is possible for morally motivated women in patriarchal societies and contexts to thrive, at least to some degree, while acting virtuously. If so, what kinds of psychologies and circumstances best contribute to women's virtuous flourishing and eudaemonic growth when confronted with and constrained by oppressive social conditions?

We take an interdisciplinary approach that incorporates research in both philosophy and psychology to answer these questions. Because virtues develop, or fail to develop, within nonidealized socially influenced circumstances of everyday life, it can be instructional to examine life stories rich with details of day-to-day challenges and struggles. In these selected life stories, self-described motives for virtues serve as narrative themes. These themes not only link abstract notions of virtue to lived actions but also link personal ideals of virtue to cultural master narratives of a good life and cultural stereotypes of virtuous men versus virtuous women. Here we focus on the life stories of and challenges faced by two women with narrative themes of contributing to society by helping and caring for others. We attend particularly to the master-narrative scripts for virtuous women and the types of gendered life circumstances that influence or constrain women's abilities to live virtuous lives and successfully carry out moral projects.

We begin by identifying and critiquing some of the contemporary American cultural ideals for virtuous women. We then discuss some of the nonideal gendered contexts and circumstances that can prevent or constrain morally motivated women from flourishing. Next, we focus on the life narratives of two women in particular to illustrate some of the ways that virtues are constrained or facilitated by their particular social ecologies and gendered circumstances. Finally, we offer realistic norms for less-than-perfect virtuous women in less-than-ideal circumstances.

Cultural Ideals of Virtuous Women

In the United States, as in most countries, there are a number of cultural master narratives on how virtuous or good women are supposed to behave. If we conduct a web search on "virtuous women" here in the United States, we are led primarily to contemporary Christian websites. These websites usually focus on the "virtuous woman" as described in the Bible's Old Testament book of Proverbs, including that "the price of a virtuous woman is far above rubies."[1] According to Proverbs, the virtuous woman is modest, submits to her husband, desires children, gives to the poor, is wise, "knows God's Word," tirelessly runs the household, and doesn't lament all that she has to do.[2] Men are warned that women who complain or express anger are especially to be avoided. In fact, "it is better to live alone in the wilderness than to live with a contentious and angry woman."[3]

Of course, not all women in the United States are Christian, and not all contemporary Christian women are willing to submit uncomplainingly to men while bearing and caring for their children. In addition, this view of the virtuous woman applies primarily to white women. Still, the Christian master narrative on virtuous women permeates American culture, and it is difficult for many women to subvert this narrative. One best-selling Christian writer, Mandy Hale, does try to resist to some degree by encouraging her female readers to love God but also to emphasize their own well being and happiness rather than focusing exclusively on the well-being and happiness of their male partners. This may mean that virtuous women leave rather than submit to abusive men and find happiness even if single.[4]

Hale's best-known quotations drawn from her writings, including from her best-selling book *The Single Woman: Life, Love, and a Dash of Sass*, have been made into an extensive and popular compilation of posters available online.[5] The sayings on these posters include:

> It's not selfish to love yourself, take care of yourself, and make your happiness a priority. It's necessary.

> Sometimes you have to move on without certain people. If they're meant to be in your life, they'll catch up.

Although Hale encourages Christian women to be strong and to leave rather than submit to abusive men, she also advocates that women avoid actively challenging, arguing with, or confronting those who are making their lives miserable. Additional posters offered on her site include, for example,

> The less you respond to rude, critical, argumentative people, the more peaceful your life will become.

> Strong women don't play victim, don't make themselves look pitiful, and don't point fingers. They stand and they deal.

In other words, Christian women should be tough but should not complain; nor should they make accusations against their abusers. And they certainly shouldn't challenge being mistreated. They should "deal" and move on. If we conduct a web search on "strong women" as opposed to "virtuous women," we see that even nondenominational master narratives encourage women to "deal" with abuse rather than actively resist it. Online posters include such quotations as "A strong woman is one who is able to smile this morning like she wasn't crying last night."[6] Strong, virtuous women are expected and encouraged to remain cheerful and silent even when suffering at the hands of others.

Another way to explore contemporary cultural narratives online about virtuous women is to search for images of the more secular phrase of "women who make a difference." Search results provide plenty of online posters with quotations from exemplary women. For example, one features a photo of Jane Goodall and quotes her as saying, "You cannot get through a single day without having an impact on the world around you. What you do makes a difference, and you have to decide what kind of difference you want to make."[7] Another features a photo of Michelle Obama saying, "Success isn't about how much money you make, it's about the difference you make in people's lives."[8] So, another master narrative tied to women's being virtuous is that of improving the lives of others either locally or globally.

But is this narrative compatible with the pervasive Christian narrative of virtuous women as submissive to men and noncomplaining? Even Michelle Obama, who arguably has made a significant positive difference in many people's lives, was and continues to be constrained by her "First Lady" role and the need to avoid complaining or appearing angry.

Recent debates on naturalized approaches to virtue ethics[9] center on the degree to which the situation a person finds themselves in, as opposed to their supposed virtuous traits, determines their behaviors. For example, is being helpful a trait or a situation-specific response? Those who argue that helping behavior is situation dependent point to studies such as one that shows that a person is more likely to exhibit helping behavior when their mood is elevated (e.g., after finding a dime).[10] But we are less interested in this trait versus immediate situation debate and more interested in how virtuously motivated persons exhibit, or fail to exhibit, virtuous behaviors over time while facing pervasive master narratives and gendered life circumstances. Women are not only faced with gendered cultural narratives and expectations about virtue but also faced with broader cultural contexts that enable or constrain their day-to-day virtuous behaviors. In particular, they are embedded within patriarchal cultures and patriarchal institutions, while also being held to rigid norms of femininity and masculinity.

Women's experiences are often traumatic, danger infused, and stressful. In patriarchal cultures, women are much more likely than men are to experience rape, sexual assault, sexual harassment, intimate violence, and psychological abuse. Many women also experience unwanted pregnancies, with accompanying risks of death or damage to health when there is no access to safe abortions. Or, conversely, many women experience devastating miscarriages during wanted pregnancies. Women are much more likely than men are to engage in stressful and undervalued care work for children, the elderly, those with dementia, those with special needs, and those in need of nursing care. In other words, while being barraged with cultural messages that women should develop submissive and self-diminishing virtues, many women also face nonideal, sometimes traumatizing circumstances that prevent them from effectively following through on and/or flourishing while attempting to help others or "make a difference" in society. These gendered circumstances can result in both emotional and material impediments to engaging in virtuous projects and activities or may entirely prevent women from engaging in the moral projects and activities they hold dear.

We aim to contribute to the expansion of virtue ethics theory by theorizing the development of virtues under oppressive and nonideal conditions. We solicit and examine life stories, and focus here on stories in which self-described motives for virtues serve as narrative themes that were identified by the authors. As previously noted, these themes not only link abstract notions of virtue to descriptions of lived actions but also link personal ideals of virtue to cultural master narratives of a good life.

Interdisciplinary Approach

Our approach to the study of virtue theory and moral development is interdisciplinary and focuses on how people construct an understanding of themselves and others in their life stories—an understanding that inherently involves ethics and cultural concepts of virtue.[11] Peggy DesAutels, the first author, is a philosopher who researches virtue theory, feminist ethics, and moral psychology. Jack Bauer, the second author, is a psychologist who researches narrative psychology, self-identity, and eudaemonia. Both are committed to the view that the construction and reconstruction of one's own life story is an attempt to showcase how one identifies with particular virtues and strives to put them into action, such that virtues serve as narrative themes. For instance, some life stories feature a personal importance placed on concerns such as cultivating others' welfare (e.g., via caring relationships or fighting injustices) and one's own wisdom, authenticity, and skills for engaging in virtuous activities. By using these themes and other narrative tools, individuals attempt to construct a story of a good life of their own—a good life story.[12] These stories also draw from cultural values, norms, and ideals, and they allow for quantitative and qualitative/critical analysis.

Our "Self, Motivation, and Virtue" study collected the life stories of 100 interviewees.[13] The two-to-three-hour interviews were conducted on Skype and then transcribed. Participants also filled out a one-hour survey. The life story interview questions were modified from those of McAdams to address the topic of virtue development.[14] Questions fell into categories of brief life chapters, key scenes (e.g., high points in life, low points, turning points), life challenges (e.g., loss, gender challenge), personal values and beliefs (e.g., role model, religious and political views), and personal development (notably "becoming oneself within one's family environment"). For this chapter we draw on the interviews of two particular women participants to address the following questions:

How do motives and life goals shape the development of virtues?

How does one's social ecology both foster and constrain that development?

How do difficult circumstances both thwart and facilitate virtues?

What virtues should women have in patriarchal cultures and nonideal gendered circumstances?

We selected these two women's stories for several reasons, most notably because each of their life stories had an overarching moral theme of helping others that was situated in contexts of conflicts and difficult choices that women are especially likely to face.[15] We narrow our focus on the life stories of two women who emphasize the virtues of "helping others" or "making a difference" in others' lives. In these two (white, college-educated, American) women's life stories, we can see that in spite of their reasonably privileged social and financial circumstances, there remain gendered costs to living virtuous lives. They experience a variety of constraints on following through on virtuous motivations, including difficult and even traumatic circumstances, but we also see resilience in the wake of their particular nonideal gendered circumstances. Owing to space restrictions, we present only a truncated summary of (and brief commentary on) their life stories as they illustrate the ways that gender and virtue intersect in women's life stories.

Introducing Joan and Maria

Like many we interviewed, Joan and Maria emphasized life-story themes of helping others and making a difference We selected them from among our interviewees for having virtuous motivations combined with particular life challenges and gender-based events that reflect the types of challenges faced by many of the women we talked with. At the time of their interviews, Joan was in her mid-forties and Maria was in her late forties. Both are of comfortable socioeconomic status (as we interpret their interpretation of their living conditions).

Joan grew up in the Midwest and had a self-described "normal childhood" with a domineering and authoritarian father. In college, she majored in engineering and "came to Christ" through the Ministry of the Crusade. After college, she married and worked outside the home as an engineer for three years. Her husband, like her father, was also domineering and authoritarian. After her children started arriving, she became a full-time mom and homeschooled all three of her children. During this time, her husband was without stable employment, and at her husband's insistence, they often changed churches and housing. It was also during this time that Joan's father needed cancer care, so he moved in with Joan and her family for four months and then died. More recently, she moved her mother nearby into an assisted cancer-care facility. Six months ago, Joan separated from her husband of twenty years, and she is currently looking to get back into the engineering workforce.

Maria describes herself as having a "Leave It to Beaver" idyllic childhood and teen years in a small East Coast town, where her grandfather owned a 1950s-style candy and soda shop. She moved to the Midwest to attend a Catholic college, where she took philosophy and theology courses that guided her as a person and "helped her find her spiritual path." While in college she participated in a service trip to Appalachia. This trip strengthened her desire to help others in need. During what she describes as her "reservation years," Maria worked as a nutritionist on a Native American reservation and developed very close friendships with women there. Next, she worked as a social worker at a women's shelter. During this time, she was "pressured into sex" by a male acquaintance and became pregnant. She made the heart-rending choice to have an abortion. After training as a nurse, she went "back to the reservation." While she was pursuing a graduate degree, a close friend (who had a wife and young daughters) committed suicide. She moved in with the family to be supportive. After a few months, she made plans to resume her educational path, but changed course again to care for her sick aunt back home. Her aunt died two months later, and she relocated for her residency but dropped out after only one week because her dad died unexpectedly. She spent six months helping her mother after her dad's death and was then called back again when her mother fell and broke her hip.

When asked to describe the overarching aim or theme in their lives, Joan replied that it was investing her life into other people's lives "so that they can grow and invest in other people's lives." She elaborates that this "can be in the church through discipleship, or it could be through works, through mentoring, even in homeschooling." Similarly, Maria's overarching aim is "making a difference" by "watching other people find their strength" and through community engagement.

Both Joan and Maria are committed to the Christian faith but through very different denominations. Joan is a conservative evangelical and Maria is a devout Catholic. They had very different career paths. Joan was employed as an engineer after college and then quit working outside the home to homeschool her three children. Maria ministered primarily to other women as a Catholic social worker in a women's shelter and as a nutritionist, nurse, and nurse practitioner at Native American reservations. In spite of their differences, both are deeply committed to caring for and improving the lives of others.

The self-described low point of Joan's life was having a miscarriage. The worst part about the miscarriage for Joan was that her husband failed to acknowledge its significance and just wanted to move on as if it hadn't happened. Maria's low point was having an abortion after becoming pregnant as a result of being "pressured for sex" from a male acquaintance.

Joan, currently separated from her husband, described some of the difficulties in her marriage:

> My husband and I went for counseling with the associate pastor at the church I'm attending right now. We have been to ... 8 to 10 churches in our ... 20 years of marriage.... And my husband was again saying we needed to move churches. In this counseling session, we're going over why I did not want to move churches. I just sat there and I said, "My best friend has died, my mother has [a serious illness] and is probably going to die in the next 5 years. My kids are happy here. My oldest is questioning whether she even is a Christian, and if you move her churches again, she doesn't even want to go" ... just pleading with him to just stop being difficult. And he doesn't even see it that way. He sees it that people aren't following the Bible and he's defending God. But he ended up walking out on the counselor after about six sessions. And then he sent these awful emails to him and the head pastor and to my Sunday school teacher and to the lady who was counseling me outside of that. And ... one of the guys came up to me and asked me is your husband [psychologically disordered], and I was just like, I don't know. I don't even know what that term means. Does he just go crazy and do stupid stuff? Yeah, if that's what [psychologically disordered] is, maybe, I don't know.

Why might Joan, being an independently minded engineer, stay in what many would term a psychologically abusive marriage for twenty years, obeying her husband's wishes on church and other matters even when both she and her children were harmed by his decisions? Joan notes that "at the time" her perspective was that there was "no way out [without] church discipline," because she had "made a vow before God," and "in the churches we attended, the focus was on the verse that 'God hates divorce' [except in cases of] adultery or abandonment." This scenario reflects a pernicious problem around the notion of personal agency within (in this case Christian) religious systems. On the one hand, Joan voluntarily chose this belief system. On the other hand, the belief system required her to surrender agency, even in the face of harm. It takes an act of tremendous courage and development to place such vows, which are presumed to be eternal, in a broader context and to separate from her husband.

Although Maria never married, she experienced what she termed "mystical love" with Jim (name changed). They were often in different locations over the years but kept in touch through phone calls. And although they had this special connection to each other, there was also, from Maria's perspective, "mistrust and betrayal" in the relationship. She explains:

> [Several years later] when he was aware I was leaving the reservation specifically to come back to live with him, he was already living with this other woman and didn't tell me about it.... I really felt like God played some huge trick and God got it wrong because we had been so spiritually connected ... and it took me years [to recover].... We had this unconditional love thing going. How could you [God] not have us have a future? [Jim and I] can look back at our journey together and say that we're grateful for all of it ... there were so many points that were so tumultuous ... there was mistrust and betrayal ... but ... we weren't doing them purposefully to hurt each other. We were living our lives and answering our own questions.... It's like even though those things that we did hurt each other so deeply, they were coming from a place of innocence.... Every part of our walk together has unfolded just as it should.

Each in their own way, Maria and Joan spent years committed to and loving men who mistreated or betrayed them. Joan decided to separate, but Maria forgave Jim and thinks everything "unfolded just as it should." Both of their life stories fit with the virtuous Christian woman narrative—both submitted in various ways to the men with whom they were in intimate relationships, at least for many years, and allowed these men's hurtful behaviors to go unchecked for extended periods of time.

When asked about her greatest life challenge, Joan replied that it was "critical and negative" people. She elaborates on what she means: "People who have large amounts of needs and very little ability to consider other people's perspectives. High negativity, high negativity, high criticism, strong ideas and beliefs, but very little compassion or consideration for others."

For Maria, her greatest life challenge is body image and weight: "I grew up with an eating disorder ... before the days when they called them eating disorders." She doesn't elaborate much, but we should note that a surprising number of highly accomplished and seemingly "together" women we interviewed mentioned weight or health as a greatest life challenge or conversely as an accomplishment if they had managed to lose weight and keep it off. When we listened to the recordings of the interviews without knowing what the person being interviewed looked like, we were usually surprised, based on everything else we had heard, that weight was an issue for them since they often seemed to be so self-assured and "together."

As women, Joan and Maria experienced a number of difficult circumstances that men would not have faced, or at least would have experienced much differently. We refer to these situations and events as "gendered circumstances." These gendered circumstances include interrupting careers to care for loved ones, navigating patriarchal institutions, and surviving abusive relationships.

During the interviews, we asked respondents about their own perspectives on the role that gender plays in their lives. In particular, we asked the following question of everyone we interviewed:

> Think about the single greatest challenge or problem that you have faced that is tied to being a [woman or man]. Perhaps other people have treated you unfairly or differently on account of being a [woman or man]. Perhaps you have felt pressure to live up to the ideal image of being a [woman or man]. In any case, please describe what this challenge or problem is, how the challenge or problem developed, who was involved, how you have addressed or dealt with it, and what it might say about you or your life.

Joan's answer to this question concerned the treatment of women at her church:

> So probably the hardest thing that I've dealt with [are] the [extremely conservative] churches that my husband has taken us to.... I find as a female, you have no voice. It's, of course, elder-ruled, who are all men. The husband is consulted. The wife is never really talked to and even really considered.

Maria, on the other hand, stated that "nothing comes to mind" in answer to this question and explains, "I just never found being a woman challenging." She does note that her best friend, who works in the National Park System, has gender-based challenges on the job, but points out that her own work life has mostly been women-centered.

In response to questions about their own perceived virtues and vices, Joan's self-described virtues were patience and compassion. Maria's were an acceptance of and love for others. Joan's vices were interrupting and intimidating others: "I interrupt people a lot because my husband says too much.... Sometimes I think I intimidate people [because of my engineering style of thinking and talking]." Maria's vices were a lack of initiative and procrastination: "My greatest strength is my greatest weakness in some ways because sometimes it seems like I so have gotten used to everything unfolding as it should that I don't make the efforts to make things actually happen. I guess procrastination has to be up there, too."

A prominent virtue in Joan's and Maria's life stories is generativity, which is the psychological term coined by Erik Erikson for the virtue of care, specifically in the form of contributing to future generations of society.[16] But generativity, like other virtues, does not develop as an unalloyed good.[17] Generativity is a strong theme in women's cultural master narratives, but part of that story is for women to bear a disproportionate share of that virtue.[18] Joan's and Maria's self-stated virtues—patience, compassion, acceptance, and love—fit well within the cultural master narrative of women as not only nurturing but also compliant. As for vices, Joan explains her tendencies for interrupting and intimidating by appealing to the masculine master narrative of engineering; such characteristics do not fit within the feminine master narrative of caring for others. In contrast, Maria frames her vices as the flip side of her traditionally feminine virtues, which is to say a lack of agency, even as her life story exhibits several sustained instances of agency, particularly in the form of generativity. Joan's and Maria's life stories are consonant with the life stories of other women in this same study: enacting the master narrative of generativity requires a sacrifice of some autonomous pursuits (in love, in work), even as the theme of generativity provides its own source of meaning in life.[19] While the ideal of generativity cuts across genders, women consistently report higher levels of generative concern.[20] Furthermore, the burden of sacrifice for generativity, particularly in matters of marriage and family, has been shown to fall disproportionately on women.[21]

Gender and Virtue

How should women approach their lives when they have as an overarching goal that of engaging in meaningful moral projects and contributing to the flourishing of their communities? In our less-than-ideal world, it is safe to assume that most women have survived some sort of traumatic experience (e.g., sexual assault, harassment, intimate partner violence). It is also safe to say that many women are underpaid, stressed, and at risk of burnout tied to care work and professions. The nursing and teaching professions, for example, are especially taxing. In one study, over 40 percent of hospital staff nurses scored "in the high range" for job-related burnout.[22] In another study, 93 percent of elementary school teachers experienced high levels of stress, and teachers with low levels of stress and high coping ability were "practically non-existent."[23]

A nonideal approach to virtue ethics includes an approach that emphasizes virtues needed for survival, coping, recovery, and resilience. Women must, first and foremost, pay special attention to developing themselves as effective and flourishing moral agents under oppressive and otherwise less-than-ideal circumstances. When possible, they should research and educate themselves on how best to improve the lives of others (locally and/or globally), develop social and relational skills, develop interpersonal support networks, engage in meaningful moral projects, and live out their passions. Even so, conflicts are inevitable. Even though generative behavior yields a heightened sense of well-being for the giver,[24] caring for others and caring for the self often conflict.[25]

A virtuously motivated woman's actual life story most likely includes responding as best she can to oppressive and constraining circumstances while minimizing damage to her moral agency and her own flourishing. This could include resisting domination, abuse, and harassment in her personal and work life, resisting harmful gender-based cultural stereotypes that lower a sense of self-esteem and self-worth, taking steps to ensure her personal safety, and recovering from such traumas as intimate violence, rape, sexual assault, and harassment. Striking this balance between growth and safety is not easy; it generally requires a high level of perspective-taking, pragmatic experience, and psychosocial maturity.[26]

When possible, women should develop and maintain caring relationships that contribute to their flourishing. These relationships could include intimate relationships, family relationships, and close friendships. They could also include broader-reaching caring relationships for others in their local and global communities. A virtuous woman in nonideal relationships and stress-inducing caregiving circumstances does her best to cope with or extricate herself from these relationships and circumstances. Nonetheless, women are often unable to avoid stressful care work for loved ones with illness, dementia, or special needs. They may even be unable to avoid care work for difficult, angry, or even abusive loved ones, and they may be unable to leave harmful or abusive relationships. At the very least, they may need to give up the moral projects about which they are passionate to engage in unchosen and stressful care work or remain in challenging and harm-inducing family or intimate relationships.

Such accounts of nonideal virtuous lives would include how narrators educate themselves on how to improve the lives of others (locally and globally), find opportunities to improve their social and relational skills, and develop support networks. Unfortunately, not all who are motivated to live virtuous lives flourish while attempting to do so. For most of us, virtuous lives are difficult and stress filled. As Kathryn J. Norlock puts it:

> We are better off with the heavy knowledge that evils recur than we are with idealizations of progress, perfection, and completeness, and if we cultivate an appropriate ethic for living with such heavy knowledge, it should not prevent us from doing our best to resist evils, improve the lives of victims, and enjoy ourselves. Instead of hoping for something like Rawls's "realistic utopia," we can cultivate the skills of more resilient knowers in a nonideal world constituted by perpetual struggle.[27]

A virtuous woman in nonideal circumstances copes and flourishes as best she can under the circumstances. She knows her strengths, capabilities, and limits. She is caring, but she is also mindful of the persistence of injustices and evils in spite of one's best efforts to improve the lives of others, and she settles for less than perfection in the struggle to cultivate her own virtues.

Notes

1. Proverbs 31:10.

2. Proverbs 31:10–31.

3. Proverbs 21:19.

4. Hale, *The Single Woman.*

5. These can be found by conducting a web search for images of "Mandy Hale quotes."

6. This particular quotation is often attributed to Harriet Morgan. See, for example, https://quotefancy.com/quote/1640302/Harriet-Morgan-A-strong-woman-is-someone-who-is-able-to-smile-this-morning-like-she-wasn.

7. See, for example, https://www.azquotes.com/author/5682-Jane_Goodall.

8. See, for example, https://www.azquotes.com/quote/543494.

9. See Flanagan, *Varieties of Moral Personality*; Harmon, Moral philosophy meets social psychology; Merritt, Virtue ethics and situationist personality psychology; Doris, *Lack of Character.*

10. Doris, *Lack of Character*, 30–31.

11. McAdams, *The Redemptive Self*; Taylor, *Sources of the Self.*

12. Bauer, Eudaimonic growth.

13. While space restrictions limit our description of the methods, we note that the 100 participants are alumni of a midwestern university and were each paid $100 for the interview and an accompanying personality survey. For more details on the project and our methods, see Bauer and DesAutels, "Toward an integrated psychology and philosophy of good life stories." This research project was partially funded by the John Templeton Foundation (Templeton Religious Trust) through the Self, Motivation, and Virtue Project.

14. McAdams, *The Stories We Live By*.

15. Each woman was interviewed separately via Skype at a single sitting by the second author. The interviews lasted between two and three hours. Their and others' names have been changed here for confidentiality.

16. McAdams and Guo, Narrating the generative life.

17. Swanton, Developmental virtue ethics.

18. Stewart and Vandewater, "If I had it to do over again ..."

19. Bauer and DesAutels, When life gets in the way.

20. See, for example, Hart et al., Generativity and social involvement among African-American and among White adults.

21. Stewart and Vandewater, "If I had it to do over again ..."; Tessman, *Burdened Virtues.*

22. Pearlin et al., Caregiving and the stress process.

23. Herman, Hickmon-Rosa, and Reinke, Empirically derived profiles of teacher stress, burnout, self-efficacy, and coping and associated student outcomes.

24. McAdams et al., Stories of commitment.

25. Bauer and DesAutels, When life gets in the way.

26. Ibid.; Cox et al., Generativity, the big five, and psychosocial adaptation in midlife adults.

27. Norlock, Perpetual struggle, 7.

Bibliography

Bauer, Jack J. Eudaimonic growth: The development of the goods in personhood (or: cultivating a good life story). In *Handbook of Eudaimonic Well-Being*, edited by J. Vittersø, 147–174. Cham, Switzerland: Springer, 2016.

Bauer, Jack J., and DesAutels, Peggy. Toward an integrated psychology and philosophy of good life stories. In *Self, Motivation, and Virtue: New Findings from Philosophy and Psychology*, edited by N. E. Snow and D. Narvaez. New York: Routledge, in press.

Bauer, Jack J., and DesAutels, Peggy. When life gets in the way: Generativity and the development of non-idealized virtues in women's life stories. *Journal of Moral Education* 48, no. 1 (2018): 126–145.

Cox, K. S., Wilt, J., Olson, B., and McAdams, D. P. Generativity, the big five, and psychosocial adaptation in midlife adults. *Journal of Personality* 78 (2010): 1185–1208.

Doris, J. M. *Lack of Character: Personality and Moral Behavior*. New York: Cambridge University Press, 2002.

Flanagan, Owen. *The Geography of Morals: Varieties of Moral Possibility*. Oxford: Oxford University Press, 2016.

Flanagan, Owen. *Varieties of Moral Personality: Ethics and Psychological Realism*. Cambridge, MA: Harvard University Press, 1991.

Hale, M. *The Single Woman: Life, Love, and a Dash of Sass*. Nashville, TN: HarperCollins, 2013.

Harmon, G. Moral philosophy meets social psychology: Virtue ethics and the fundamental attribution error. *Proceedings of the Aristotelian Society* 99 (1999): 315–351.

Hart, H. M., McAdams, D. P., Hirsch, B. J., and Bauer, J. J. Generativity and social involvement among African-Americans and White adults. *Journal of Research in Personality* 35 (2001): 208–230.

Herman, K. C., Hickmon-Rosa, J., and Reinke, W. M. Empirically derived profiles of teacher stress, burnout, self-efficacy, and coping and associated student outcomes. *Journal of Positive Behavior Interventions* 20, no. 2 (2018): 90–100.

McAdams, D. P. *The Stories We Live By: Personal Myths and the Making of the Self*. New York: Morrow, 1993.

McAdams, D. P., Diamond, A., de St. Aubin, E., and Mansfield, E. D. Stories of commitment: The psychosocial construction of generative lives. *Journal of Personality and Social Psychology* 72 (1997): 678–694.

McAdams, D. P., and Guo, J. Narrating the generative life. *Psychological Science* 26 (2015): 475–483.

McAdams, Dan P. *The Redemptive Self: Stories Americans Live By*. New York: Oxford University Press, 2013.

Merritt, M. Virtue ethics and situationist personality psychology. *Ethical Theory and Practice* 3 (2000): 365–383.

Norlock, Kathryn J. Perpetual struggle. *Hypatia* 34, no. 1 (Winter 2019): 6–19. https://doi.org/10.1111/hypa.12452.

Pearlin, L. I., Mullan, J. T., Semplem, S. J., and Skaff, M. M. Caregiving and the stress process: An overview of concepts and their measures. *The Gerontologist* 30, no. 5 (1990): 583–594.

Stewart, A. J., and Vandewater, E. A. "If I had it to do over again ...": Midlife review, midcourse corrections, and women's well-being in midlife. *Journal of Personality and Social Psychology* 76 (1999): 270–283.

Swanton, C. Developmental virtue ethics. In *Developing the Virtues: Integrating Perspectives*, edited by J. Annas, D. Narvaez, and N. E. Snow, 116–134. New York: Oxford University Press, 2016.

Taylor, Charles. *Sources of the Self: The Making of the Modern Identity*. Cambridge, MA: Harvard University Press, 1989.

Tessman, Lisa. *Burdened Virtues: Virtue Ethics for Liberatory Struggles*. New York: Oxford University Press, 2005.

7 Since Morgenbesser: The Continuing Messiness of the Social Sciences and Some Related Incoherences

Alasdair MacIntyre

It has been my very good fortune to have been a member of the same philosophy department as Owen Flanagan three times—at Boston University in the early 1970s, at Wellesley College in the early 1980s, and then at Duke University in the 1990s—and I am deeply grateful for how much I was able to learn from him as a result. During that long period, neither he nor I ever made the philosophy of the natural and social sciences the principal focus of our attention, but neither of us could avoid extended reflection on some issues raised in those areas. I now return to such issues, focusing initially on two related topics: more generally on the relationship of our understanding of the natural sciences, especially physics, to our understanding of the social sciences, especially the kind of sociology with which I was once engaged; and more particularly on the defeated aspiration to provide a compelling philosophical justification for the claim that social science is and must be reducible to natural science, and on why, at least so far as that kind of sociology is concerned, that defeat was inevitable.

A presupposition of this chapter is nonetheless the greatness of the achievements of the Vienna Circle and their heirs. It may indeed seem later that in some respects I remain too respectful of their positivism. And it is certainly true that, were I to tell the story of the journey away from that positivism in the past sixty years, I would draw on the resources of those who have stressed the continuities of later thought in the philosophy of science with positivism. Thomas Kuhn, I am persuaded, was Carnap's heir as well as critic.

Kuhn is one of three key figures in my narrative. The two others are Sidney Morgenbesser and British sociologist Tom Burns. Kuhn has been written about a great deal too much, and as a result his true importance is often obscured. Morgenbesser has been written about a good deal too little, and as a result his importance has too often gone unacknowledged.[1]

In 1970, while I was making the transition from the Sociology Department at the University of Essex to the Philosophy Department at Boston University (with a short stay at Brandeis University in between), Dorothy Emmet and I published a collection of what we took to be the most important recent papers by philosophers and sociologists on the nature of social scientific, especially sociological, understanding and theorizing.[2] We included Morgenbesser's 1966 article "Is It a Science?" because it seemed to us a crucial intervention in what turned out to be the penultimate stage of the debate between, on the one hand, those who held that the findings and theories of the social sciences, rightly understood, had or could be given the same form, the same logical structure, as the findings and theories of the natural sciences, understood as the heirs of the Vienna Circle had understood them, and, on the other hand, those who denied this. There were of course deniers of various kinds—phenomenologists, Wittgensteinians, and others—and there was more than one version of the positivist view. Morgenbesser, who wrote from within the positivist tradition, proceeded by constructing what he called "an ideal type of social scientific naturalism," one strongly indebted to Ernest Nagel's formulations, and therefore closely akin to Hempel's and Carnap's, noting that this "canonical version of social scientific naturalism maintains that all the basic theories of the social sciences ... are reducible to some suitable theory in the natural sciences."[3]

Morgenbesser's contention was that neither party to this controversy had been able to make anything like good on its claims, that neither side had the philosophical resources that would have justified its proponents in concluding in the one case that such a type of reduction is impossible and in the other that there are no significant obstacles to achieving it. And here Morgenbesser's anxiety not to advance one inch further than he took himself to be entitled to by his argument may conceal from his readers the importance of the conclusion that he was on the edge of reaching.

This is often so in Morgenbesser's writings, although it was not at all so with his conversations. He is now most often remembered for his philosophical jokes (see, for example, his Wikipedia entry) and for the spontaneous production in discussion of telling counterexamples. Sometimes, as in his instant refutation of J. L. Austin's incautious generalization about double affirmations ("Yeah yeah"), joke and counterexample were one and the same. Both jokes and counterexamples can be conversation stoppers, but with Morgenbesser they were more often conversation starters. The joke or counterexample redirected and refocused attention by suggesting some further and unexpected line of thought. So it was sometimes also in his articles. In the case of "Is It a Science?," Morgenbesser's argument that in the philosophical controversy over reduction neither side had been able to defeat the other suggested a possibility hitherto not sufficiently entertained: that the question of whether the statements of the social sciences, or at least of this or that particular social scientific theory or explanation, are reducible to statements in the natural sciences—or at least to this or that particular set of statements in some particular natural science—is just *not* the kind of philosophical question that either the positivists or their critics and opponents took it to be. Questions about reduction have a clear sense *within* physics and some other sciences, and such scientists have local standards that enable them to say whether some particular reduction has been achieved or whether they have significant evidence for the possibility of some other reduction. But where no actual research inquiry has yet been proposed, the answers to questions posed in the abstract about possibilities of large-scale reduction at some time in an indefinite future become elusive, and denials of such possibilities will be as misleading as assertions. The kind of frustration that Morgenbesser identified is not after all surprising.

The continuing history of debates about reduction—and about emergence—in the intervening decades confirms this. We have numerous examples of scientific projects that give a clear sense to what is to be understood as a successful reduction in this or that particular instance. But nothing in this longer history has any significant bearing on the philosophical question of whether sociology should be structured as a natural science is structured.[4]

So in what direction should we have then moved? It was Morgenbesser who suggested how to move beyond Morgenbesser. I begin from what he says about defining science by turning from theories to theorists, to the "procedures and collective aims" of scientists: "Modern science is distinguished by ... institutionalized methods of communication of results and criticism, methods which have resulted ... in the miracle of modern science, men building on the labor of others."[5] And elsewhere Morgenbesser was to say of the scientific community that "its members share higher-order attitudes about beliefs and attitudes to revise them in the light of the objections to them brought by others."[6]

In such passages, was Morgenbesser perhaps anticipating the later move away from positivist philosophy of science to sociological and even social constructivist theories, according to which it is the contingencies of external social life that to some large extent determine the paths taken by scientists? To think so would be to miss his point. What Morgenbesser believed was that the shared commitments of scientists were constituted by the norms of their science, by their shared acceptance of the authority (1) of a given body of theory understood as having a certain structure and (2) of a set of rules as to when and how elements of that body of theory may be revised or rejected and perhaps replaced. Scientists are those whose higher-order attitudes are such that they are in fact committed to excluding from influence, as far as possible, precisely those external factors on which social constructivists place so much emphasis. To think and act as a scientist is to think and act such that retrospectively one's work can be understood in terms that are internal to science.

The paradigmatic examples of sciences so understood, if Morgenbesser was right, are of course physics and chemistry. But, before we raise further questions, we might remind ourselves of some important autobiographical remarks that had been made by Kuhn about his experiences during a year (1958–1959) at the Center for Advanced Studies in the Behavioral Sciences: "I was struck by the number of the overt disagreements between social scientists about the nature of legitimate scientific problems and methods.… The practice of astronomy, physics, chemistry, or biology normally fails to evoke the controversies over fundamentals that today often seem evident among, say, psychologists or sociologists."[7] That is to say, the community of, for example, sociologists differed notably from the community of physicists in that they lacked precisely that agreement in their shared higher-order attitudes toward their own beliefs and attitudes that Morgenbesser took to characterize genuine scientific commitment and genuine scientific community.

There was not among sociologists, as there was among physicists, a single shared conception of the overall enterprise to which they were contributing and of the rules and values that they must respect if they were to be recognized as contributing to that enterprise. It is not that sociologists did not have conceptions of the overall enterprise in which they and their colleagues are participating, but the plural matters. There were and there still are a number of competing conceptions, each giving expression to its own view of how inquiry should be directed. Such conceptions certainly have in common a certain range of overlapping subject matters. And their proponents coexist happily or unhappily within—within what? I almost said "the same discipline," but the unity of the discipline seems to consist in not very much more than that provided by the bureaucratic organization of university departments and professional associations.

More than forty years ago, in 1976, Geoffrey Hawthorn published a remarkable history of sociology, a history that badly needs to be provided with a sequel covering the years since 1970 and therefore such remarkable contributions as Goffmann's account of role-playing, Garfinkel's ethnomethodology, Philip Rieff's analysis of secularized culture, and the sociologies of religion of Bryan Wilson, David Martin, R. Stephen Warner, and Christian Smith, along with a myriad of detailed studies of various topics. But the view of sociology in all its heterogeneity provided by such a sequel would coincide with Hawthorne's in one important respect. Hawthorne said that sociology "has not consisted, as a Kuhnian model might suggest, of the periodic overthrow of 'paradigms' after long periods of relatively settled 'normal science.'"[8] And what was then true remains true. Indeed, none of Kuhn's key insights seem to find application to the history of sociology, yet sociologists themselves have often believed, and some continue to believe, otherwise.

That explains the tragicomic aspect of some sociologists' unrequited love affair with Kuhn. Consider, for example, Robert W. Friedrichs's once notable *A Sociology of Sociology* (1970), in which paradigms proliferate, without any reference to what was involved, in Kuhn's account, in such paradigm cases of paradigm shifts as those from impetus theory to Galilean and Newtonian physics or from Newtonian mechanics to either quantum or relativistic mechanics. What Kuhn—and others—had liberated sociology from, according to the story told by Friedrichs, was "the predominance of logical positivism in the philosophical thinking of natural and social scientists in America during the 'forties and 'fifties."[9] But that story obscured two central aspects of Kuhn's work that at first sight seem in tension with each other. The first is the extent of the continuity of Kuhn's theorizing with that of his positivist—or, if you like, logical empiricist—predecessors, especially Carnap. The second is the genuinely radical implication of Kuhn's theses about incommensurability and untranslatability.

On the continuity with Carnap, the best statement is by Gürol Irzik and Teo Grünberg. Irzik and Grünberg do of course recognize important differences between Carnap and Kuhn. Whereas, for example, Carnap held that the correspondence rules of a theory both give empirical meaning to theoretical terms and provide application for the theory to phenomena, Kuhn argued that, although sometimes this latter function was discharged by correspondence rules, the application of theory to phenomena was, in crucial cases, a matter of an ability to see and identify key similarity relations, an ability exercised in the light afforded by particular examples.[10] But such differences have the significance that they have only against a background of underlying agreements, and Irzik and Grünberg were able to conclude "that Kuhnian normal science corresponds to activity within Carnapian theory" and that what for Kuhn is revolution as paradigm shift is for Carnap "a transition from one theory to another precisely when the theoretical postulates change."[11] And they were able to make good on this claim in some detail.

Even where there are salient differences between Carnap and Kuhn, Irzik and Grünberg are able to show that there is more continuity than Kuhn himself had always recognized, so that, although Kuhn acknowledged that Carnap had preceded him in recognizing that there were alternative languages in which a set of theories might be expressed, he still ascribed to Carnap the view that "for anything properly classified as scientific knowledge it could be both stated and scrutinized in either language, using the same method and getting the same result."[12] Any reader of Carnap's *An Introduction to the Philosophy of Science* (1966) would have taken this to be Carnap's view, yet Irzik and Grünberg are able to show that Carnap had in fact anticipated Kuhn in making a greater allowance for the possibility of untranslatability between the languages in which theories are stated than Kuhn had recognized.[13]

Kuhn was of course right in taking himself to have understood in a quite new way the crucial part played by the problems of translation between conceptually and theoretically incommensurable languages in the history of physics. What is required of physicists is not only that they learn to speak such different languages but also that they become able to move between the idiosyncrasies of those languages so they can speak in one language about what can be said—or said accurately—only in some other. To be educated as a physicist involves being initiated into the fluent use of such languages as they inform experimental and observational practice, as well as in theorizing, and this at once sharply distinguishes the education of physicists from that of sociologists. To explain why, let me sketch all too briefly what is involved in the practice of sociology, or rather of one particular kind of sociology, for sociology only comes to us in kinds.

The particular sociology of which I am going to speak is that of Tom Burns, especially in his earlier research. I choose it partly because of its outstanding sociological interest and partly because of its logically instructive qualities. And that earlier research dates from the same period in which Morgenbesser and Kuhn were at work. Writing about the project that provided the data for his book *The Management of Innovation* (1961), coauthored with G. M. Stalker, Burns described two important discoveries that he made (in 1967).

The first was the result of his asking a number of managers in the firms that he was studying to give him an account of their career in the firm. These accounts would be "lucid, well-organized and informative" but would stop short at the time at which the managers had arrived at their present position. When asked about their ongoing activities, they would explain what they and their department had been doing in the past and would be doing again quite soon, when the present crisis or unusual rush of work or reorganization was over and they and their colleagues had returned to normality. It turns out that in many types of organizations, those who play key roles have a conception of normal functioning as something that was in place until quite recently and will be in place again quite soon but that just happens for exceptional reasons to be absent now. So those who participate in such organizations are deceived and deceive themselves as to the improvisational character of much of their rule-following and therefore as to the rules themselves.

In addition, when Burns compared the records made by those managers of meetings between them and their subordinates with the records of the same meetings made by those subordinates, it turned out that on a large number of occasions the managers had been misinterpreted by their subordinates, a misinterpretation of which both parties had remained unaware even though it informed their subsequent actions and interactions. Add further that when Burns had those managers keep diaries in which each noted how much time he spent on different types of tasks, they and he discovered that the proportion of their work time spent on their central management tasks was significantly less than they had hitherto reported and believed.

These three discoveries were in fact incidental and unexpected outcomes of a research project aimed at distinguishing those features of organizational life that facilitate genuine innovation from those that obstruct it, but by themselves they afford valuable matter for generalizations about illusions concerning the normal and the exceptional, conversational misinterpretation, misconceptions of the working day, and the part that these illusions, misinterpretations, and misconceptions may play in the lives of organizations. What kinds of generalizations are they? Let us call them "characteristically and for the most part generalizations," generalizations that have the same properties as Aristotle's "*hōs epi to polu*" generalizations.

They are not the universal lawlike generalizations of the physicist. They are not the kinds of probabilistic generalizations that hold for fundamental particles. Given a loosely defined set of conditions under which such generalizations can be expected to hold—under which it is always well worth asking whether in this or that particular case they do in fact hold—there will always remain an indefinite range of possible exceptions, exceptions that characteristically lead us to modify our view of the scope of our generalizations rather than to treat them as counterexamples warranting a rejection of this or that generalization.

Such generalizations find their place within larger accounts of the type of organization that Burns was studying: engineering firms in Scotland about to enter what was then a new market for electronically controlled devices. What was required of such firms was a capacity for change and innovation—for change and innovation as a new way of corporate life—and Burns's task as a sociologist was to identify the characteristics of firms able to transform themselves in this way from firms unable to do so. As it turned out, this was most of the firms involved. (One crucial variable was the relationship between, on the one hand, the career structure of a firm and its effect on the goals of individuals at different stages in their careers and, on the other hand, the goals of the firm.) What emerged was a set of cautious and limited generalizations whose importance was in part to have suggested questions about how other types of organizations respond in similar situations. The direction of inquiry was from particular case to particular case, not toward the construction of some overarching theory. (I might add that the sad history of the construction of overarching theories in sociology itself provides justification for proceeding as Burns did.)

It might seem then that in so characterizing these sociological generalizations I have already provided some justification for speaking, as I do in the title of this chapter, about the messiness of the social sciences as compared with physics or chemistry. But there is more to this messiness, for if we want to explain what happens in the life of some organization, such as those studied by Burns, we will have to invoke not only the relevant set of generalizations but also from time to time the impact on that organization of historically contingent external factors. And we possess no set of reliable generalizations that govern both the ongoing life of organizations and the factors that impinge on them from without, no generalizations analogous to those laws that govern both, say, the movements of some planetary system and the forces exerted by some body impacting it from outside.

Both the messiness and the complexity can be brought out by considering what it would be to provide a sociological explanation of the life and death of the management of that extraordinary hedge fund of twenty years ago, Long-Term Capital Management, whose collapse in 1997 for a moment endangered the global financial system.[14] There would have to be at least three levels of explanations involved.

First, the initial success of the fund is to be ascribed to the reliance of its traders on what they took to be the truth of the Efficient Market Hypothesis and on the predictive power of equations that enabled them to evaluate the prices of derivatives and calculate offsetting risks. Those formulas were derived from the Black-Scholes method for estimating the correct price of a stock future, and Nobel Prize–winning economist Myron Scholes was himself a partner in Long-Term Capital Management. But it is important that we distinguish between the amount of the traders' activities that was governed by the relevant sets of interpreted equations and the amount that was governed by their seemingly uncritical confidence in the predictive power of those equations and by their well-developed ability to ignore signs of predictive failure. It is at this point that we need exactly the type of generalization formulated by Burns, a generalization that would explain aspects of the traders' activities by appealing to the part played by their misconceptions of the normal and the exceptional and by their misinterpretations of others.

We do not, as far as I know, possess any extended sociological study of the relevant kind. But some clues as to what such a study might disclose are to be found in autobiographical accounts of activities in such firms.[15] What we do know is that the activities of those at Long-Term Capital Management required what looks in retrospect like an unconsciously willed unawareness of the possible impact of external factors on the markets on whose predicted stabilities the traders relied, a remarkable lack of any sense of historical contingency. The particular history to which those traders were blind was that of the Russian economy, and it was the unpredicted collapse of that economy in 1997 and the unpredicted reactions to that collapse by the Russian government that fatally undermined the strategies of the traders at Long-Term Capital Management.

Donald Mackenzie, in a pioneering study of the sociology of arbitrage, trading that profits from discrepancies in asset prices, as exemplified in the trading practices of Long-Term Capital Management, has argued that what needs to be explained is its vulnerability to the effects of the Russian crisis. He has an insightful explanation of that vulnerability: "LTCM's success led to widespread imitation," and the effect of that widespread imitation was in the end "a cascade of self-reinforcing adverse price movements."[16] But, as Mackenzie recognizes, this explanation also requires a lack of awareness by traders of their self-exposure to risk.

So, to provide anything like an adequate explanation, we have to bring together a set of economic equations precise in their formulation but with a less than precise range of applications, a set of sociological generalizations with exactly the characteristics that I described earlier, and two historical narratives, one of the development in the relevant period of the economic environment in which Long-Term Capital Management at first flourished and one of the history of the relationship between government and markets in post-Soviet Russia. The overall explanation of the collapse of Long-Term Capital Management would therefore involve contributions from three academic disciplines and appeal not only to two kinds of generalizations but also to a complex historical narrative. This is characteristic of many social scientific explanations, so any attempt to offer a general account of explanation in the social sciences will have to take into consideration this kind of complexity, which is quite alien to physics or other natural sciences.

Nonetheless, such sociological, sociological-cum-historical, or sociological-cumeconomic-cum-historical generalizations have genuine explanatory power. There is a limit to such explanatory power, which is reached when certain types of highly structured, norm-governed, goal-directed activities are encountered. These types of activities are well designed to exclude the influence of exactly those factors that sociological explanation invokes. Consider, for example, the game of chess. Sociology may be able to explain many things about those social structures and those cultural norms that explain whether chess flourishes as an activity. There is nothing to stop there being both a history and a sociology of chess clubs or of chess tournaments, and the growth in number and popularity of chess tournaments in England after 1851 has been used to explain the new attention to opening strategies by chess players. But sociologists can have nothing to say concerning the problems consequently confronted chess players; the solutions to those problems that were proposed and accepted, rejected, or revised; the new set of problems that were then posed; and the types of reasoning put to work in perfecting the relevant strategies. There is of course a history of chess, in part a history of the changing place of changing opening strategies in the game. But the psychological or sociological motivations that particular chess players may have had are irrelevant to, and throw no light on, those strategies in their play that were advances beyond those of their predecessors and an example for their successors, thus securing for them a place in the history of chess.

As with chess, so it is with physics. There can be sociological explanations of research institutes and teaching departments, of why resources for work in a particular area in physics were available but not for work in some other area. What there cannot be is a sociology of, say, the developments in physics narrated in Kuhn's *Black-Body Theory and the Quantum Discontinuity, 1894–1912*, or of those that culminated in Richard Feynman's work in quantum electrodynamics. What kind of "cannot" is this? In both chess and physics, the participant contributes only insofar as she or he is guided in devising strategies, in constructing experiments, or, in theorizing, by what are not just taken to be but are the best standards known so far to either chess players or physicists for acting, arguing, and judging as they do. And explanation, insofar as it is explanation of the chess or the physics, is explanation in the light of those standards, explanation that terminates with those standards. There may be sociological or psychological explanations of why this or that chess player or physicist failed to have regard for such standards, of why, for example, some particular group of physicists engaged instead in what Feynman called "cargo cult science." But to answer the question of why the standards were what they were would be simply to recapitulate the history of the chess or the physics. More than this explanation, sociological or otherwise, cannot be achieved.

Such a "cannot" is not, of course, the conclusion of an a priori argument. As we might have learned from Morgenbesser, no a priori argument on such matters will yield either a "can" or a "cannot" of any interest. But we only have to try to imagine generalizations purporting to identify sufficient psychological or sociological conditions for such occurrences as, say, the crisis in quantum mechanics in 1924–1925 to recognize that we have no idea at all of what it would be to formulate such a generalization with any plausibility. Helge Kragh, in his excellent social history of physics, treats some claims of this kind with perhaps more seriousness than they deserve, but in a way that brings out their shortcomings admirably.[17] He therefore rejects, for good empirical reasons, Paul Forman's claim that the "socio-ideological circumstances" of the Weimar republic predisposed German physicists toward an "acausal quantum mechanics," because they "craved for a crisis in the existing semi-mechanical atomic theory."[18] It may be more apt to put it like this: any such claim is going to sound indistinguishable from an Alan Sokal parody.

We can now see how the earlier aspiration to reduce the social sciences to the natural sciences and the later project of understanding the natural sciences in social constructivist terms had something important in common. Both were blind to the radical heterogeneity of our modes of inquiry and understanding. If we list those characteristics that justify us in calling physics a science, then plainly sociology is not by those standards a science. But to say that sociology is not by those standards a science is not in the least to denigrate it.

I suggested earlier that one way to understand the heterogeneity of the disciplines better is to contrast what it is to be educated as a physicist, to be initiated into physics, with what it is to be educated as a sociologist, to be initiated into sociology. Begin with physics. What is involved in becoming a physicist? At the outset students have to move from their everyday life-world into a systematically Newtonian universe in which what is invariant is not weight but mass, and in which a well-defined concept of mass has displaced an earlier unclear, but for a long time familiar, concept of quantity of matter. What was and is mass, so conceived? It is, as Richard S. Westfall summarized Newton, "an activity of resistance evoked in changes of state," something that "establishes the equation between an impressed force and the change of motion it produces."[19]

To have become nowadays an inhabitant of a universe conceptualized in Newtonian terms is to have encountered (1) a set of problems already solved long ago, such as, for example, that of the relationship of inertial mass to gravitational mass; and (2) a set of problems that are initially statable in, but not solvable in, Newtonian terms, problems whose solution required the introduction of a radically new vocabulary. So the concept of mass is transformed when Newton's laws are replaced by Einstein's and mass is no longer treated as invariant. Students learn to redefine and to rethink what they mean by "mass," something brought home to them even more strikingly when their movement out of Newtonian mechanics is not into relativistic mechanics but instead into quantum mechanics, so they become able to entertain thoughts such as those of a particle with zero mass, but with spin.

Such transitions in vocabulary—from an everyday to a Newtonian and then from a Newtonian to a relativistic or quantum-mechanical standpoint—are not just a matter of changes in the meanings of words that we use. They are, as David Bohm emphasized, shifts in our habits of thought, breakdowns of established habits and the acquisition of new ones.[20] But unlike some other changes in habits, the acquisition of these new habits does not result in anything like the disappearance of the old ones. No matter how sophisticated we become in our relativistic or quantum-mechanical ways, we find that we cannot avoid remaining in many contexts Newtonians and in some others even pre-Newtonians. We will have learned to think in several new ways that are at odds both with the others and with the habits of our everyday life-world.

To understand adequately what is involved in such learning, we need to note another crucial dimension of a contemporary education in physics. We grasp the new concepts that we acquire only insofar as we grasp sets of equations. And our progress is toward increasingly advanced mathematics, as we move from Cauchy-Riemann equations to Lie algebras and beyond, so that the mathematical differences between the initial life-world of students and the quantum-mechanical or relativistic universes that they learn to inhabit can only be appreciated retrospectively. It is the absence of the relevant mathematics from most popularizing books about physics that misleads their readers, concealing the nature of the gap between the world as understood by physics and the everyday life-world inhabited both by the readers of such books and by physicists themselves in their everyday lives.

In both these respects, the education of a sociologist is very different. There are indeed a multitude of concepts specific to sociology, but these invoke an extension of, rather than a break with, concepts already in use among educated people. Indeed, some of these sociological concepts have enriched our shared vocabulary: Weber's *charisma*, Durkheim's *anomie*, Bourdieu's *habitus*. And even with those concepts where happily this is not the case, as with the heterogeneous networks of Bruno Latour and John Law, there is nothing particularly striking about learning how to use them, certainly no movement between incommensurable standpoints or partially untranslatable vocabularies, no rupture in our linguistic habits.

As for the mathematics involved in sociology, it has a number of uses: the statistical analysis of data, the use of mathematical models to chart social relationships and progressions, including Markov chain modeling, applications of graph theory in identifying structures, and applications of path analysis in identifying causal relationships. Much of the mathematics put to those uses is also put to use in physics, but a crucial difference is that the findings of the sociologist can be communicated independently of the mathematics without significant loss of content. There may be some rare exceptions to this, but I do not know them.

What sociologists who study organizations do have to acquire are some highly specific skills that enable them to carry out tasks of three kinds: tasks of description, of explanation, and of the achievement of understanding.[21] What is required by way of description is not initially the sociologist's own descriptions but the descriptions of their status, role, and function provided by those who are the objects of study. So Burns, both in his earlier inquiries about engineering firms and in his later studies of the British Broadcasting Corporation (BBC), conducted a large number of lengthy interviews. Only the initial questions in those interviews were formulated in advance. What mattered was to allow the earlier responses to guide the later questions, so that the self-disclosures of the interviewees could take directions unpredicted by the questioner and unexpected by the interviewees.

To the skills of the interviewer there have to be added the skills of identifying in the material so provided inconsistencies, tensions, and signs of conflict, of the unacknowledged, and of the anomalous. And whatever explanations are given for them have to be integrated into a larger scheme of explanation that enables us to account for, and to assign causal responsibility for, the success or failure of the organization in achieving its goals. What makes something a factor in the explanation of such success or failure varies with context, so in the case of Burns's earlier study of engineering firms, one crucial factor in their success or failure in innovating was, as I already noted, the career structure of that industry in that time and place. The career goals of some individual managers led them to act in ways that obstructed innovation by their firms. Yet, in Burns's first study of the BBC, an organization in which there was at that time a marked preoccupation with career paths, it turned out that this was a period of brilliant and innovative achievement in television production. The emphasis on career paths seems to have had few, if any, negative effects. Why?

To answer this question, we have to consider the final goal of sociological inquiry, as Burns understood it, the achievement of understanding through description and explanation. Understanding involves interpretation of the speech and activity of the participants in an organization, so that we understand both "the structure and functioning of the organization" and "the values, motives, and purposes of the people within it." Two conditions must be satisfied by a successful interpretation. It must enable us to relate the first of these to the second and it must "square with" the understanding of their own activities by the participants. Sociological understanding both extends and corrects, sometimes radically corrects, the understanding of participants.[22]

In the case of the BBC, what mattered was that talented individuals of a number of different kinds had come to understand how they attended to their own concerns by serving the purposes of their art, so that "engineers, scene-shifters, directors, actors, stage-managers, porters, lighting supervisors, accountants, cameramen, secretaries and so on" knew what they had to do. They had learned that they "had to dovetail in with what other people were doing, or had just done, in a meticulous and precisely timed way."[23] Burns compares their activity in which "commands, directions or prompts are hardly ever necessary" and "emergencies ... are signals for the immediate performance of appropriate and complementary tasks by all the members of the team" with "what, at its superlative best, one finds happening in an operating theatre between surgeon, anaesthetist and theatre nurses" and with the performances of "fishing crews and ensembles of actors or acrobats or musicians."[24]

What happens in such a situation is that the organizational forms serve the ends and purposes of the relevant practice, of the relevant activity, rather than the activity being organized to serve the functions of the organization, the functions as understood by those who are administrators or managers rather than participants engaged in the relevant practice. So Burns's conclusion about this productive activity in the BBC was that "none of it seemed to be *managed*."[25]

What management of such organizations can do at best is no more than to provide the kind of institutional setting in which constructive activity aimed at the achievement of some good internal to that activity can take place. What management cannot do is to plan and control that activity, to manage it, imposing their mandates on the actors. What management in fact often does is *either* to insist on managing such activity, thereby frustrating it, *or*, when the activity flourishes independently of their attempts to plan and control it, to take undeserved credit for the outcome, thus sometimes deceiving others and most often deceiving themselves.

To have achieved understanding of an institution or organization, then, is to have identified the relationships between the values, actions, and purposes of the individuals who compose it and the structure and functions of the institution or organization and to have understood those relationships in a way that characteristically the participants themselves could not understand them, since it is their failures to understand that are often crucial determinants of institutional or organizational outcomes. (Never forget that in some cases misunderstanding is necessary for organizational success.) Yet the language involved in expressing this kind of understanding is essentially one already spoken by the participants.

What physics trades in is theory. What sociology trades in is a grasp of this set of organizations or that set of institutions in this or that time or place. And those whose activities constitute the particular organization or institution that is the sociologist's object of study can, if instructed by the sociologist's findings, achieve new understanding of themselves in their roles and activities, a new self-knowledge that is at odds with and that would be apt to undermine the currently dominant self-image of the effective manager. Unsurprisingly, one outcome is a series of devices that enable managers to avoid or resist the acquisition of self-knowledge, so the relationship of sociology to practice is quite different from the relationship between, say, physical theory and engineering or architectural practice.

Were I able to develop the contrast between physics and sociology further, I would want to focus on this relationship between theorizing and practical activity. But instead I need to make one final point.

My central aim has been to emphasize the heterogeneity of the disciplines of inquiry as exemplified in the large differences between physics and sociology, but there is more to this contrast than a catalog of differences. What physics provides is an account of the universe according to which there are, among other things, photons and leptons, but no physicists and no sociologists either. It is a portrait of a universe from which the possibility of causally effective intentional activity is not merely absent but has been excluded. Yet both physicists and sociologists cannot avoid understanding themselves as being engaged in precisely such causally effective intentional activity, and without such activity sociologists of course would have nothing to study. Thus our understanding of ourselves and of nature is fractured. What one discipline tells us, another denies.

Philosophers have offered an intriguing range of possible solutions to this problem, but as so often happens in philosophy, they have had and still have too many solutions to offer, each incompatible with the others, each unable to defeat its rivals except on its own terms. We therefore have naturalistic and antinaturalistic solutions, physicalist and emergentist solutions, and champions of and critics of supervenience, with the result that we have been provided with no more than a set of promissory notes. In our everyday practice, we learn to live with incoherence by putting it out of mind. How we do this and more generally how we negotiate incoherences in our everyday social lives is, I have suggested, something that sociology has been instructing us about for quite some time. That this is one of its distinctive achievements has, however, gone largely unnoticed by philosophers of the social and natural sciences. This, too, is a fact that deserves further attention.

Notes

1. Although see Cauman et al., *How Many Questions?*, especially the introduction by Jonathan Lieberson.

2. One paper dated from 1943, one from 1953, and the rest from the 1960s. See Emmet and MacIntyre, *Sociological Theory and Philosophical Analysis*.

3. Morgenbesser, Is it a science?

4. For an excellent account of what has been at stake in that history, see Gillett, *Reduction and Emergence in Science and Philosophy.*

5. Morgenbesser Is it a science?, 27.

6. Morgenbesser, Is it a science?

7. Kuhn, *The Structure of Scientific Revolutions*, xiii.

8. Hawthorne, *Enlightenment and Despair*, 243.

9. Friedrichs, *A Sociology of Sociology*, 94.

10. Kuhn, *The Structure of Scientific Revolutions*, 198–202; Irzik and Grünberg, Carnap and Kuhn, 287–298.

11. Carnap, *An Introduction to the Philosophy of Science*, 298.

12. Kuhn, Afterword.

13. Irzik and Grünberg, Carnap and Kuhn, 296–301.

14. For a narrative account, see Lowenstein, *When Genius Failed.*

15. See, for example, Knee, *The Accidental Investment Banker.*

16. Mackenzie, Long-Term Capital Management and the sociology of arbitrage, 353.

17. Kragh, *Quantum Generations.*

18. Forman, Scientific internationalism and the Weimar physicists, 153.

19. Westfall, *Never at Rest*, 419.

20. Bohm, *The Special Theory of Relativity*, 114.

21. Burns, Sociological explanation. For an account of Burns's work and importance, see Poggi, Tom Burns.

22. Burns, introduction to *Description, Explanation and Understanding*, 15–16.

23. Ibid., 17.

24. Ibid., 18.

25. Ibid., 17.

Bibliography

Bohm, David. *The Special Theory of Relativity*. New York: W. A. Benjamin, 1965.

Burns, Tom. *Description, Explanation and Understanding*. Edinburgh: Edinburgh University Press, 1995.

Burns, Tom. Sociological explanation (1967). In *Sociological Theory and Philosophical Analysis*, edited by Dorothy Emmet and Alasdair MacIntyre, 55–75. London: Macmillan, 1970.

Burns, Tom, and Stalker, G. M. *The Management of Innovation*. London: Tavistock Books, 1961.

Carnap, Rudolf. *An Introduction to the Philosophy of Science*. New York: Basic Books, 1966.

Cauman, L. S., Levi, I., Parsons, C. D., and Schwarz, R., eds. *How Many Questions? Essays in Honor of Sidney Morgenbesser*. Indianapolis, IN: Hackett, 1983.

Emmet, Dorothy, and MacIntyre, Alasdair, eds. *Sociological Theory and Philosophical Analysis*. London: Macmillan, 1970.

Forman, Paul. Scientific internationalism and the Weimar physicists: The ideology and its manipulation in Germany after World War I. *Isis* 64, no. 2 (January 1973): 150–180.

Friedrichs, Robert. *A Sociology of Sociology*. New York: Free Press, 1970.

Gillett, Carl. *Reduction and Emergence in Science and Philosophy*. Cambridge: Cambridge University Press, 2016.

Hawthorne, G. *Enlightenment and Despair: A History of Sociology*. Cambridge: Cambridge University Press, 1976.

Irzik, Gürol, and Grünberg, Teo. Carnap and Kuhn: Arch enemies or close allies? *British Journal of the Philosophy of Science* 46 (1995): 285–307.

Knee, Jonathan. *The Accidental Investment Banker*. Oxford: Oxford University Press, 2006.

Kragh, Helge. *Quantum Generations*. Princeton, NJ: Princeton University Press, 1999.

Kuhn, Thomas. Afterword. In *World Changes: Thomas Kuhn and the Nature of Science*, edited by Paul Horwich, 313–314. Cambridge, MA: MIT Press, 1993.

Kuhn, Thomas. *The Structure of Scientific Revolutions*. 2nd ed. Chicago: University of Chicago Press, 1970.

Lieberson, Jonathan. Introduction. In *How Many Questions? Essays in Honor of Sidney Morgenbesser*, edited by L. S. Cauman, I. Levi, C. D. Parsons, and R. Schwarz, 1–14. Indianapolis, IN: Hackett, 1983.

Lowenstein, Roger. *When Genius Failed: The Rise and Fall of Long-Term Capital Management*. New York: Random House, 2000.

Mackenzie, Donald. Long-Term Capital Management and the sociology of arbitrage. *Economy and Society* 32, no. 3 (2003): 349–380.

Morgenbesser, Sidney. Is it a science? (1966). In *Sociological Theory and Philosophical Analysis*, edited by D. Emmett and A. Macintyre, 20–35. London: Macmillan, 1970.

Morgenbesser, Sidney. The realist-instrumentalist controversy. In *Philosophy, Science, and Method: Essays in Honor of Ernest Nagel*, edited by S. Morgenbesser, P. Suppes, and M. White, 200–218. New York: St. Martin's Press, 1969.

Poggi, Giannfranco. Tom Burns, 1913–2001. *Proceedings of the British Academy* 120 (2003): 43–64.

Westfall, R. *Never at Rest: A Biography of Isaac Newton*. Cambridge: Cambridge University Press, 1980.

8 Why Forgiveness Is So Elusive

David B. Wong

Introduction

This chapter is an investigation into forgiveness that is in the spirit of much of Owen Flanagan's work, especially *The Geography of Morals*. It is a broadly empirical and comparative investigation that is meant to stand in contrast with the more typical treatment accorded to important moral concepts. As Flanagan characterizes it, the more typical treatment is justification of "ways of thinking and being that we already assume and prefer."[1] There is little thought about how wide-ranging comparative investigation could change the very way we conceive of the conceptual phenomena to be explained. Flanagan writes,

> Moral philosophy as an academic subdiscipline, especially in the precincts that are considered most rigorous, still favors and selects for Rubik's Cube–type minds over the sorts of historical and anthropological curiosity and critical political sensibility that are valued across most of the humanities, in many sectors in the human sciences, and in the adjacent territories of social and political philosophy (e.g., in critical race theory and feminist philosophy).[2]

I dedicate this chapter to Owen, my friend and colleague, with whom I have ranged through these territories.[3]

There are at least a couple of reasons why forgiveness is an entirely appropriate subject for the kind of investigation Flanagan calls for. First, it is not unusual for philosophers to suppose that it is universal or to treat it as if it were, again, without much informed critical reflection on what "it" could be. P. F. Strawson, for example, in his hugely influential work on the reactive attitudes, takes it to be one of the cognitive-affective-conative attitudes that are part of human nature.[4] Second, the question of whether forgiveness is a good thing relates to Flanagan's provocative questioning of the necessity for any form of anger or almost all of it.[5] If, as Strawson characterizes it, forgiveness is letting go of resentment, and if resentment is a form of anger or a near cousin to it, then letting go of anger might be letting go of the need for forgiveness. In the *Geography of Morals*, Flanagan classifies resentment as seeming to be, or likely to be, in the anger family.[6] He points to moral traditions that have envisioned the elimination of anger or that do not recognize anger as we do in WEIRD (Western, educated, industrialized, rich, and democratic) societies. If anger or resentment is not regarded as inevitable or desirable, then forgiveness might have no honored place in the list of moral emotions. Or perhaps the "it" of forgiveness might be more variable and elusive than supposed in contemporary Anglo-American philosophy, and the question of whether "it" has a place might be much more complex. This is where I have ended up in my investigation, so far. Let me show you why.

The Cultural and Historical Character of Forgiveness: Early Greek, Chinese, and Buddhist Philosophy

To get a tentative fix on forgiveness, consider more specifically Strawson's characterization: "To ask to be forgiven is in part to acknowledge that the attitude displayed in our actions was such as might properly be resented and in part to repudiate that attitude for the future (or at least for the immediate future); and to forgive is to accept the repudiation and to forswear the resentment."[7] Resentment, says Strawson, is "a reaction to injury or indifference, and reactive attitudes in general rest on, and reflect, an expectation of, and demand for, the manifestation of a certain degree of goodwill or regard on the part of other human beings towards ourselves; or at least on the expectation of, and demand for, an absence of the manifestation of indifferent disregard."[8] Because Strawson describes forgiveness as something like an exchange between forgiver and forgiven, let me call this conception of forgiveness the "interactive" model.

In his book *Before Forgiveness*, David Konstan makes a strong case for concluding that interactive forgiveness, which he also identifies as "modern," "interpersonal," or "dyadic" forgiveness, did not exist among the ancient Greeks and Romans. The reason is that the ancients were not so much interested in repudiation of attitude or "an inward change of character as a condition for reconciliation with a wrongdoer."[9] Konstan adds that "the appeasement of anger and the relinquishing of revenge were rather perceived as resting on the restoration of the dignity of the injured party, whether through compensation or gestures of deference, or else by way of discounting the offense on the grounds that it was in some sense involuntary or unintentional,"[10] and that virtue that might be displayed by the person in power is clemency, which is "not so much forgiveness but rather gentleness or mildness," which is letting the offender "off as a special grant of generosity."[11]

In line with this characterization of the ancient Greek and Roman eras, Aristotle mentions in the *Nicomachean Ethics* that *sungnômê*—the Greek term most commonly rendered as "forgiveness"—is appropriate when people act either under external compulsion or in excusable ignorance of the facts or circumstances.[12] What does the use of *sungnômê* mean here? Konstan draws the implication that a harmful or inappropriate act that was performed involuntarily is excusable or understandable and does not count as a case of wrongdoing for which the agent is responsible.[13] He argues that in cases of this kind we do not apply the term *forgive* but rather words such as *exculpate* or *absolve*.[14]

What might explain the absence of interactive forgiveness? Charles Griswold points to the perfectionistic structure of Greek and Roman ethics. It is focused on articulating and recommending the character of a person of complete virtue. Such a person would have no need of forgiveness, and he would regard himself as largely immune to being harmed morally, with the exception of being deprived of warranted honor.[15] The way for those who seek to reconcile themselves with Aristotle's *megalopsuchos*, or great-souled person, is through the practices of compensation, deference, or presentation of excuses. Of course, these practices are quite different from one another, and which one might be appropriate can depend on the nature of the offense or perhaps the wish of the offended.

Perhaps a perfectionist ethic is not such a congenial home to interpersonal forgiveness. Such is Kwong-loi Shun's conclusion about Confucian ethics. To ask whether forgiveness is recognized in Confucian ethics, Shun stipulates a certain meaning of resentment: that it is a response of "asserting ourselves in light of wrongful treatment that we regard as challenging our standing."[16] The standing concerns our deserving to be treated in certain ways and with proper respect. In this way, resentment can be conceived of as an assertion in defense of one's self-respect. One grants forgiveness if one deems that the offender has shown sufficient remorse and resolution to change for the better with regard to how he has treated us.

Shun systematically runs through terms in classical Chinese that live in the semantic neighborhood of interactive forgiveness, and he argues that no convergence in meaning is to be found. For example, *she* (赦) refers to someone's refraining from doing something negative to another person when he has the authority or power to do so, such as the ruler of a state who frees an official of an enemy state who has been captured in war. There is no psychological dimension of coming to see the offender in a different light but rather a public act of refraining from doing something negative to another person.[17] The word *kuan* (寬) does refer to a change in mind-set that can lead to lenient action, but the mind-set is not primarily on the offending party; it is rather on correcting a deficiency in the self.[18] The change might be from a narrow-minded view of the offender to a broad-minded view.

This last point about *kuan* segues nicely into Shun's explanation of why forgiveness is absent in Confucian ethics. It has to do with the Confucian view of what is truly disgraceful. In ancient China, it was commonly assumed that being insulted leads to disgrace unless one fights back or counters the insulting behavior in some other way. To address the pervasive fighting that resulted from this assumption, philosopher Songzi proposed that one should simply stop viewing insults as a disgrace. Though he doubted that this change would reduce fighting, the Confucian Xunzi proposed that the truly disgraceful should not be tied to the way others view us but be a matter of our own ethical conduct, which includes how we respond to others' treatment of ourselves.[19] This view is broadly shared across the Confucian tradition. In *Analects* 14.1, Confucius is asked about *chi* (恥), which is usually translated as "shame" but should be read with the connotation of being tainted. Confucius says that when the *dao* (道) prevails in a state, take [public] office. To take office when the *dao* (道) does *not* prevail is to *chi* (恥), to be tainted. In *Mengzi* 2A2, a lower form of courage is described as fighting in response to insulting treatment; the higher form is to act on what is ethically appropriate.

As Shun puts it, "When one is wrongfully injured the focus of one's attention should be on the ethical quality of one's response rather than how one's standing is challenged by the offender."[20] One's standing should be a matter of one's own ethical quality rather than the way one is viewed or treated by others. Shun thinks that interactive forgiveness is based on ethical standards specifying what is due to oneself and not allowing treatment in violation of those standards. The absence of interactive forgiveness in Confucianism is based on ethical standards specifying what one's qualities and actions should be.[21] The first set of standards legitimates resentment, at least within certain bounds. The second bids us to rise above resentment.

There is one passage that might be taken to run against Shun's interpretation of the thinking behind the lack of a Confucian concept of forgiveness. That is in *Analects* 14.34. The implications that passage has for Shun's interpretation, however, are difficult to figure, in great part because of difficulties in translation and interpretation. In that passage, Confucius is asked whether to repay *yuan* (怨) with *de* (德). Confucius replies by asking what, then, would one repay *de* (德) with? He goes on to say, "Repay *yuan* [怨] with *zhi* [直] and repay *de* [德] with *de* [德]."

The meaning of this passage depends on several controversial choices one must make in translating it. Suppose we translate *de* (德) as "virtue," the most common meaning it has in the text as a whole. Suppose we translate *yuan* (怨) as resentment, with the meanings discussed earlier. Suppose we translate *zhi* (直) as "straightforwardness," a meaning that is compatible with the one it appears to have in *Analects* 5.24. Then the passage is rendered in the following way: "Someone said, 'To employ virtue to repay resentment, how would that be?' The Master said, '"What, then, would you employ to repay virtue? Employ straightforwardness to repay resentment; employ virtue to repay virtue.'"[22]

Translating *de* (德) as virtue raises awkward questions: "Why *wouldn't* you repay anything directed towards you with virtue? Are we directed not to be virtuous in certain circumstances?" This is why other translators have rendered *de* (德) in one of its other meanings, as "kindness."[23] Contemporary philosopher Yong Huang, in his guide to the *Analects*, renders it as "a good turn."[24] The translation of *yuan* (怨) as "resentment" also makes for puzzles. Confucius's reply to the question does not consider whether one has done anything to deserve resentment, which one would think quite relevant to how one should repay *yuan* (怨). The alternative translation of Slingerland[25] and Huang of "injury" makes more sense in the context.[26] Finally, translating *zhi* (直) as "straightforwardness" raises puzzles also. It has the connotation of not hiding or being deceptive about one's true intent or feelings, but this simply raises the question of what one's intent or feelings should be. If Confucius is simply saying that one should express whatever is on one's mind, this is a remarkably lax piece of moral advice. A preferable translation of *zhí* (直) is "uprightness."[27]

The upshot is my preference for the following translation: "Someone said, 'To repay injury with kindness (or a good turn), how would that be?' The Master said, 'What, then, would you repay kindness (or a good turn) with? Repay injury with uprightness, and repay kindness (or a good turn) with kindness (or a good turn).'"

The remaining important question is the meaning of "uprightness." In *Analects* 17.24, Zigong says that he doesn't like the type of person who mistakes exposing people's wrongdoing for uprightness (and this passage seems fully compatible with Shun's interpretation that the Confucian attitude toward wrongdoing done to oneself is more concerned with one's own ethical conduct). In the *Analects*, uprightness is generally associated with doing the right thing and, in one place, with making others upright.[28] If the idea is to respond to injury in a manner that is right or appropriate for the particular case, which is the basic meaning of rightness or *yi* (義), and in such a manner as to encourage the other person to do what is right, then I take it that Confucius's advice is not any particular course of action but to repay injury in such a way as to be appropriate for the occasion and so as to encourage the other to do what is right in the future. Thus interpreted, *Analects* 14.34 not only makes sense in Confucian terms but also is compatible with the spirit of Shun's interpretation.

There is one other wrinkle to consider in the Confucian attitude toward resentment toward the family. In *Mengzi* 6B3, resentment is given an uncharacteristically positive role, at least in a restricted context.[29] In that passage, Mengzi says about a chapter in the *Shijing* (詩經) that it is not the complaint of a petty person but the expression of the *yuan* (怨) of a man toward his parents, who have committed a major wrong (though *yuan* is not appropriate when expressed toward a minor wrong committed by parents). Here it may not be inappropriate to translate *yuan* as resentment, but perhaps with an asterisk. The reason is that Mengzi goes on to say that *yuan* in this context is an expression of familial *ren*, or care (親親仁也). To lack sufficient concern for parents when they have committed a major wrong toward oneself, he says, is to become even more alienated from them. Not to care about this is to be a bad child. Note that the focus is ultimately on the parents' moral standing and only secondarily on the child's, and that is why it has to be resentment* in question, because the most focus is not on what parents have done to oneself but on what they have done to damage their own moral standing.[30] The passage does seem notable for its positive regard for resentment in the specified context, but given the distinctive focus of resentment*, it also is compatible with a forward-looking ethical stance that is more focused on what can be done to improve, in this case, the offender's moral standing.

Compare this forward-looking attitude in Confucian perfectionist ethics with Griswold's reading of Greek and Roman perfectionist ethics. Resentment, the aforementioned exception aside, is not an ethically appropriate attitude for a person of complete virtue. Shun's statement of the alternative focus of the person who has been wrongfully or unjustly treated highlights the intent to be morally constructive in a forward-looking fashion, and this focus has a somewhat different direction than the one that Griswold's reading of the Greek term reveals. One should focus on responding appropriately to the wrongful act, which may not involve addressing the offender at all if there are more morally constructive acts to be done. If the offender poses more of a moral threat to others in the future, for instance, one might focus on preventing that rather than seeking some form of redress for what was done to oneself. If the offender might be influenced toward a better course in the future, the virtuous person might engage with the offender, but not necessarily in order to seek expressions of one's superiority relative to the offender or, in contrast to interactive forgiveness, to obtain expressions of the offender's remorse in relation to what she has done to herself.

The view that our moral standing is a matter of our own ethical qualities rather than the attitudes expressed in the way others treat us requires an impressive immunity to a feeling of personal injury. Charles Griswold thinks this immunity is appropriate for a sage who has met a very high bar, but it is far above common life. Either that or one would have to be insensitive to injury for reasons such as self-deception or emotional blockage.[31] Shun points out that Confucianism simply does not accept this limitation on human possibility (and I would infer that Flanagan might not either).

I, too, am on the side of not accepting the limitation. It is not that I think we should increase our estimates such that most people would become sages. Forgiveness that is not conditioned on the wrongdoer's repentance, such as the act of forgiveness by Steven McDonald, the New York City police officer who in 1986 was shot three times and permanently paralyzed by a teenager he was investigating for bicycle theft, can arise from more particular circumstances and not necessarily from a person's sagelike or nearly sagelike character. In other words, I suspect that Griswold's "realism" underestimates the capacity for at least some people we could encounter in daily life, who fall well below the lofty ideal of sage but succeed in rising far above expectations when subjected to grievous injury. Six months after McDonald was shot, on the day his assailant was sentenced to ten years in prison and also the day of his infant son's christening, his wife read a public statement on his behalf. He said,

> On some days, when I am not feeling very well, I can get angry. But I have realized that anger is a wasted emotion, and that I have to remember why I became a police officer. I'm sometimes angry at the teen-age boy who shot me. But more often I feel sorry for him. I only hope that he can turn his life into helping and not hurting people.... I forgive him and hope that he can find peace and purpose in his life.[32]

McDonald did not insist on remorse or a change of heart from the young man who shot him as a *condition* of his forgiveness. Instead, he was set on a constructive response to what happened to him. He spoke at schools about the importance of forgiveness and traveled to Northern Ireland and the Middle East to speak about reconciliation. He also addressed police roll calls, speaking about officer safety but also reminding officers "to always treat everyone they encountered on patrol with the same level of respect and kindness they'd afford their own, closest friends."[33]

McDonald's response to what happened to him was indeed more focused on the ethical qualities he desired for himself ("I have to remember why I became a police officer") rather than his mistreatment by another. This part of what he did resonates with Shun's Confucian who is focused on how she is to respond ethically to wrongdoing, not on what the wrongdoer has done to her. But McDonald said that he was *forgiving* the teenager who shot him. This does not mean that he ceased recognizing that the teenager did something wrong. It did seem to mean that McDonald could no longer see a point in harboring ill feeling toward him. What McDonald did does not fit the interactive model of forgiveness, but it would take a sterner legislator of meaning than I am to deem his declaration of forgiveness an inappropriate use of words or as referring to something lesser or other than forgiveness. How many of us without a philosopher's axe to grind would say that McDonald did not know the meaning of forgiveness? Thus, one lesson I want to draw from Shun's discussion is that the absence of forgiveness in the Confucian tradition is the absence of a specific conception of forgiveness, and that not all commonly recognized and acceptable uses of the word *forgive* conform to that conception. The ethical impulse to look forward that Shun identifies as the reason why forgiveness is lacking in the Confucian tradition is in fact found in some forms of forgiveness in the Western tradition.

Consider again the interpretation recommended for *Analects* 14.34, where injury should be repaid with uprightness rather than with kindness. Perhaps McDonald's wish that the young man who grievously injured him find peace in his life can be construed as kindness, but his wish that the young man be able to turn his life into helping rather than hurting is the forward thinking of trying to make the other upright.

Another point that emerges from the McDonald example is that we needn't think of him as a sage or near-sage to understand how he might have forgiven the teenager without condition of repentance. Very shortly, I will give examples of others who have given the kind of unconditional forgiveness that McDonald gave, and it is striking in these cases that people forgave someone who had killed a person they loved. Without being sages, ordinary people can rise to the occasion of great loss and tragedy, seeking a constructive, forward-looking response to what has happened. It was, I suggest, precisely because they have received a tremendous blow, something that threatened to devastate the meaning and purpose of their lives, that they sought new meaning and purpose in coping with what had happened. McDonald found it in advocating reconciliation of long-standing conflicts fueled by resentment and hatred. He found that this was a better alternative than maintaining corrosive anger at his assailant. Ordinary people, not just sages, can do this sort of thing, I suggest, precisely because it is lifesaving. This is not to deny that they might have had prior beliefs that could have justified their acts of forgiveness, but it is to understand why they might have been able to overcome whatever emotions of anger and aggrievement stood in the way of acting on those beliefs.

There is a position that is intermediate between saying that forgiveness unconditioned on repentance is not forgiveness and saying that it comes as one of the acceptable meanings of the term. Griswold says that interactive forgiveness (together with other conditions he lays down) is *paradigmatically* interactive.[34] This is to make room for "nonparadigmatic" forms of forgiveness, and McDonald's forgiveness of his assailant fits under that category. But what does it mean to say that a use of *forgive* is nonparadigmatic? Griswold says that the nonparadigmatic usages are "departures from and conceptually dependent upon the paradigm."[35] There is a question as to what being a "departure" from a conceptual form consists in. It has to mean something more than simply being in the minority of usage cases. Does it mean that it is "somewhat incorrect" to use the word *forgive* in that way, but not enough to disqualify it as a case of forgiveness in some sense? Without further clarification of what incorrectness amounts to, this move seems to be merely conferring a prejudice in favor of interactive forgiveness, but in a milder form. Does it mean that it is somehow confusing to use the word *forgive* in that fashion? I detect no confusion in what McDonald did. The middle ground of "somewhat incorrect" is vanishingly small. Is a nonparadigmatic usage "conceptually dependent" on the paradigmatic usage? It is so only in the sense that understanding the notion of forgoing resentment *unconditionally* on repentance implies understanding the contrast condition of forgoing resentment *conditionally* on repentance. But this is equally true the other way around![36] I find no solid ground for erecting a hierarchy that orders forms of forgiveness.

Construing cases like McDonald's in the manner suggested has the effect of softening the contrast Shun suggests.[37] In some of the cases that are fairly termed instances of forgiveness, there is an impulse to look forward that seems to override a demand to make up for the past. That impulse looks pretty much like the one to which Shun appeals in explaining the lack of interactive forgiveness in Confucianism. If we look more widely at Chinese thought but still in the classical age, we see some further points of resemblance. There is a moment in the Daoist text *Zhuangzi* that conveys a forward-looking response with acknowledgment of one's wrongdoing. Compensating for past wrongdoing takes the form of trying to find and develop in oneself what is worth saving. The text playfully conveys this in the form of a rebuke to Confucius, who is portrayed in this passage as taking an unappealingly self-righteous attitude:

> There was an ex-con in Lu, named Toeless Shushan, whose feet had been mutilated as a punishment. He heeled his way over to see Confucius, who said to him, "You were careless in your past behavior and thus have ended up in this condition. Isn't it a little late to come to me now?" Toeless said, "I just didn't

> understand my duties and undervalued my own body, and so I now lack a foot, but I come to you with something worth more than a foot still intact. Heaven covers all things. Earth supports all things. I used to think that you, Sir, were just like Heaven and earth—I never imagined you would instead say something like this!" Confucius said, "It was rude of me. Won't you please come in and teach me what you've learned?" But Toeless departed. Confucius then said, "Learn from this, my disciples! For Toeless is a one-footed ex-convict, but he still endeavors to learn, so as to make up for the ugliness of his past behavior.[38]

There is a hint of interactive forgiveness in Confucius's remark that Toeless seeks to make up for his past behavior, but again the dominant theme in this passage is the urge to look forward, and it is a kind of self-righteous backward-looking moralism that is the target.

The forward-looking element of some forms of forgiveness also resonates with Buddhist traditions. While Charles Hallissey rightly demurs from pronouncing on the presence or absence of forgiveness in Buddhism per se, given its global nature and resulting linguistic and conceptual diversity, he finds that in general Buddhist traditions highly value two distinct elements: the removal of an expectation of retribution, and renunciation of anger or resentment toward someone who has offended you. Forgiveness, on Hallissey's understanding, involves only the latter. In the Sinhala language used by Buddhists in Sri Lanka, *sama venava* and its cognate *ksamava* can be translated as "forbearance," which Hallissey characterizes as more inclusive than forgiveness. It encompasses both enduring an action done against oneself and renouncing anger or resentment toward the offender.[39]

The reason for both elements of forbearance concerns the aim of Buddhism to end all suffering. It is to desist from causing more suffering for oneself and for others by "reacting to offenses in comparable ways."[40] In a verse from the *Dhammapada*, an anthology of sayings attributed to the Buddha, it is said that in those who harbor thoughts of what was done to them by others, anger is never stilled and hatred never ceases by hatred. Loving kindness is what brings suffering to an end.[41] Embodied in this verse is the idea not only that the anger causes one to harm others but that it also harms the person who was wronged. It might enmesh her in a cycle of hatred and retribution, and the experience of such anger is internally corrosive.[42]

This self-benefiting aspect of forbearance is brought out by a Tibetan Buddhist text that says:

> At the outset the Bodhisattva observes the benefit in forbearance. He thinks, "The forbearing person in the future will have few enemies, will have few discords, will have much pleasure and satisfaction. That one will be without regret at the time of death; and with the break-up of that one's body will be born among the gods in the good destiny, the heaven world."[43]

Buddhism can generally be said to contain a concept of forbearance that incorporates something like forgiveness even if it is not to be equated with forgiveness. The forgiveness in question, however, is not interactive but rather a forward-looking overcoming of anger and resentment for the sake of oneself and others. There is no requirement of repentance on the part of the offender. Hallissey points out that *anukampa* and *dayava* can be translated as "compassion" or "pity," virtues through which one empathizes with the suffering of the offender, even as the criminal is on the way to execution, or even toward one's own would-be murderer.[44] Another passage from the same work tells the story of an ascetic who was tortured by an enraged king to test his dedication to forbearance. The ascetic, even as body part after body part is lopped off the king, feels only sorrow for the king.[45] Though much more extreme and pure as an example of forward-looking forgiveness, it resonates with McDonald's empathy for his attacker.

Religious Roots of Forgiveness in the Jewish and Christian Traditions

There is a religious history behind McDonald's use of *forgive,* so let me turn to that part of the story. According to Konstan, the modern idea of forgiveness evolved from a religious conception of sin and conversion. To illustrate this conception, Konstan discusses a text entitled *The Life of Adam and Eve,* of Jewish or Christian origin, dating somewhere between ca. 1 BCE and ca. 7 CE.[46] The crucial events begin long after their ejection from the Garden of Eden for having tasted the forbidden fruit, when Adam, at 930 years old, is at the point of death. He bids Eve and their son Seth to weep and beg God to have pity on him. They do, and Eve repeatedly declares her own terrible sin, revealing a profound realization of their culpability, insisting on her own fault and expressing remorse. God takes pity on Adam and raises him to the third Heaven to wait for Judgment Day. He yields to Eve's supplication to be buried next to Adam, even though she is unworthy and sinful. Thus not only is there repentance and remorse from Adam and Eve but a change of heart on the part of God, who had been stern in ejecting them from the Garden despite the angels' pleading on their behalf.[47] Atonement in Jewish theology as developed by the rabbis of the Talmud is very much in line with this pattern of remorse and forgiveness.[48] But, Konstan argues, we still have not arrived at the full modern conception of forgiveness, "for the focus on the Jewish people's relationship to God not only has the consequence that it is for the most part God who forgives, rather than human beings, but also that the kind of offense that requires forgiveness is a generalized rejection of the Lord, as opposed to particular wrongs committed against a fellow being."[49]

However it might have been at the beginning of the tradition, it is clear that forgiveness came to obtain between human beings, with some interesting features based on its being a restorative relationship. The offender must explicitly seek forgiveness from the injured party, even if they have already given material compensation for any tangible loss inflicted. The primary responsibility for repairing the moral gap that has opened up between offender and injured party lies with the offender. Once the appropriate gesture of repentance is made, however, then it becomes the duty of the injured party to forgive.[50] The restorative nature of forgiveness comes out clearly in a discussion of Maimonides, where sins between people and sins between people and God are distinguished.[51] Even if one injures another person only in words and not in deeds, he says, the offender must pacify the other and ask for forgiveness. If not forgiven the first time, the offender should bring three friends to make the request. If refused, the offender should try a second and a third time. If the injured party still refuses, the sin now lies on their side. Maimonides goes on to say that it is forbidden for the injured party to be cruel and not be appeased. Rather, one should be quick to forgive and slow to anger. How different such a restorative conception of forgiveness seems from those interactive versions that make it entirely the option of the injured party to forgive, no matter how repentant the offender is.

To say that the Judaic tradition holds in a form of forgiveness that obtains between people is not to deny that it is related to God's relationship to people. In fact, the duty to forgive derives from the duty of Israel to draw itself closer to God by emulating God's compassion and willingness to forgive.[52]

Konstan's survey of the New Testament reveals a range of forgiveness-related phenomena. He points out that "the term most employed for forgiveness or remission of sin in the New Testament is *aphesis*, from the verb *aphiêmi*, which has the basic sense of "let go" or "dismiss."[53] There are

Jesus's words at the moment of crucifixion: "Father, forgive [*aphes*] them; for they know not [*ou gar oidasin*] what they do."[54] Konstan points out that the logic invokes the excuse of ignorance and raises the question of what there is to forgive if fault is excused. Moreover, we might wonder what the excusing condition is: that the soldiers gambling for his garments and the thieves who were next to Jesus on their own crosses and jeering him did not know he was the son of God as well as a man who could suffer and be killed? Is it rather, given that Jesus was supposed to have died for all of humankind's sins, "a sympathetic recognition of human weakness and ignorance generally," as Konstan seems to suggest at one point?[55] The kind of excusing in question here certainly goes beyond garden-variety excuse making (e.g., shooting someone because one mistakenly thinks that person is a threat to one's own life). It is rather a reference to the benighted condition of being human.

Jesus's Sermon on the Mount in Matthew 6:11 enjoins people to pray to God to "forgive [*aphes*] us our debts [*opheilêmata*], as we also have forgiven our debtors [*aphêkamen tois opheiletais*],"[56] and in Matthew 6:15, "if you do not forgive men their trespasses, neither will your Father forgive your trespasses."[57] The word for *debt* means "what is owed," and the word for *forgive* is a standard term for remission of debt. Trespass, in turn, renders the Greek *paraptôma*, literally a "misstep" or "slip." The language of the version in the Gospel of Luke mentions forgiveness of our "sins" or faults (*hamartias*): "for we ourselves forgive [*aphiomen*] everyone who is indebted to us" (Luke 11:4).[58] The language of debt and remission, Konstan claims, "is not normally an instance at all of forgiveness in the full modern sense, because it does not imply any wrongdoing on the part of the debtor: in annulling a debt, the slate is wiped clean, as though there had been no loan in the first place."[59]

On the other hand, Konstan does note that the word *trespass* means "misstep," and in Luke what is forgiven is our sins or faults. Thus, if the language of debt and remission does not normally imply wrongdoing, it seems to slide toward that implication in the versions of the Lord's Prayer in Matthew and Luke. In fact, Konstan notes that there is something like "interpersonal forgiveness" in Luke 17:3, where Jesus admonishes his disciples, "If your brother sins, rebuke him, and if he repents, forgive [*aphes*] him." And in Matthew 18:21–22, Peter asks, "Lord, how often shall my brother sin against me, and I forgive [*aphêso*] him? As many as seven times?" Jesus replies, "I do not say to you seven times, but seventy times seven."[60] In the latter passage, Konstan tells us, there is "more on the moral requirement for the victim to be reconciled with the wrongdoer than on a change of heart on the part of the offender."[61]

In his story of the emergence of the modern concept of forgiveness, Konstan is concerned with emphasizing the differences and discontinuities between the modern, interactive concept of forgiveness that will eventually emerge over time, including the differences between divine and human forgiveness. Divine forgiveness as given by Jesus can eradicate a fault or state of sinfulness in the way human mortals can cancel a debt: he requires no change of heart, nor does he need evidence of remorse.[62] This is one reason why Konstan declines to say that a fully interactive concept of forgiveness has arrived in the New Testament.

My own reading of Konstan's careful survey of the New Testament is that the "modern" concept *is* by that time emerging, but it is emerging in the company of other modes of conciliation that are not so clearly distinguishable from something many people would to this day call "forgiveness." I have already given one contemporary instance of someone describing himself as forgiving without a change of heart or remorse in the wrongdoer (McDonald). A more recent wrenching example is the forgiveness accorded by the family of one of the victims of the killings of a Bible study group in South Carolina by a white supremacist. Nadine Collier, the daughter of seventy-year-old Ethel Lance, said to the killer, Dylann Roof, "I forgive you. You took something very precious from me. I will never talk to her again. I will never, ever hold her again. But I forgive you. And have mercy on your soul."[63] This story exemplifies the unconditional forgiveness that is modeled after the forgiveness of Jesus, and I take it that this is not very unusual at all. In light of such examples, I submit that there is something suspect in a kind of retrospective legislation of the meaning of forgiveness that deems one of the forms we use today as "modern" or "paradigmatic" (in the case of Griswold) and then proceeds to see history as a kind of evolution toward that favored form. Even forms of interactive forgiveness vary significantly in acknowledging or rejecting a duty on the part of the injured party to forgive a repenting offender, and this may signal a variation in the extent to which practices of forgiveness are strongly oriented toward the function of restoring and reconciling relationships.

There is another practice of reconciliation that Konstan is reluctant to call forgiveness but that others do. Carl Reinhold Bråkenhielm (as Konstan observes) invokes a distinction between "exculpative forgiveness," where moral criticism is withdrawn, and "admissive forgiveness," where it is not.[64] Bråkenhielm cites Jesus's plea on behalf of those crucifying him as an instance of exculpative forgiveness, and this in turn makes contact with the meaning of Aristotle's *sungnômê*. Konstan says he prefers not to call it exculpative *forgiveness*.[65] Yet to this day many take Jesus's words "they know not what they do" as a model of forgiveness: they take themselves to be forgiving as Jesus would and not as arrogating something only a divine being can grant. As in the Judaic tradition, such forgiveness (though different from the Jewish practice of requiring forgiveness only after repentance from the offender) may be regarded as being derived from the requirement to draw closer to God through emulation. The father of Heather Heyer, who was killed by a white supremacist during a protest rally in Charlottesville, Virginia, announced that he forgave her killer. He said that people need to stop the hating and forgive each other, "including myself in that, forgiving the guy that did this. He don't know no better. I just think about what the Lord said on the cross."[66] Perhaps a reason why people continue to use the word *forgive* with an exculpative connotation is the difficulty of making an absolute contrast between excusing and admissive conditions; there is more of a spectrum, with the range of excusing conditions merging at some point into the widespread conditions that make us human beings the weak, shortsighted ,and foolish creatures we typically are.

The Appearance of Interactive Forgiveness

According to Konstan, interactive forgiveness first appears in a form resembling the modern idea in the middle of the seventeenth century, a signpost being Molière's *Les Fourberies de Scapin* (*The Ruses of Scapin*), in which Léandre attempts reconciliation with Scapin by pleading, "I was wrong, I admit it … , I'm as sorry as can be … , I beg your pardon with all my heart," and he offers to throw himself at the other man's feet. Only then does Scapin finally yield.[67]

Joseph Butler is notable for bringing forgiveness down from the heavens and into the context of human resentment and the desire for revenge. He is most concerned with limiting the desire for revenge and defending resentment as a natural and legitimate sentiment among us fallen creatures, though it should often be appropriately tempered by compassion. Konstan observes that it is "remarkable" how little Butler has to say about repentance or a change of heart in the offender as a condition for forgiveness.[68] It is not until we get to Kant and his championing of moral autonomy that the way is paved for conceiving moral transformation as a precondition for earning forgiveness. Repentance and the achievement of a new self are now to be judged not by God but by the person who has been wronged.[69]

As I have been suggesting, however, we ought to consider the possibility that the conceptual phenomena associated with forgiveness have a diverse, fluid, and plastic quality with no clear teleological trend toward interactive forgiveness or any specific form of it as the pinnacle. Let's go back to Butler. He characterizes forgiveness as a state reached when a victim is "affected towards the injurious person in the same way any good men, uninterested in the case, would be; if they had the same just sense, which we have the injured person to have of the fault."[70]

Perhaps there is a good reason why Butler hardly discusses repentance as a condition for forgiveness. His characterization of resentment that is appropriately tempered is in fact pretty close to the recommendation on anger that Shun attributes to the neo-Confucian Zhu Xi. To illustrate the point, Shun asks us to imagine a Confucian official unjustly slandered by rivals. The ideal attitude to be taken by the official is not "I have been unjustly treated," but rather "An injustice has been done. What am I to do about it?" The evil, says Shun, is not felt as a personal offense. The anger felt at injustice done to oneself is viewed in the same way as an injustice done to others.[71] Now, because Shun defines resentment as having an intrinsically personal element, he concludes that forgiveness, as the overcoming of resentment, is not present in Confucian ethics. Butler, on the other hand, thinks of forgiveness as including an appropriately tempered resentment, but his characterization of the result sounds much like Shun's description of the appropriate *absence* of resentment.

Butler's position illustrates a phenomenon that has popped up more than once in this very brief survey of forgiveness-related phenomena. The conceptual elements appearing in these phenomena have a kaleidoscopic quality. When the tube is twisted, the elements fall into different patterns of relationship to one another. With a twist, some elements may fall out of the picture completely, and others that have been recessive may take their place in the foreground. There is enough continuity in the elements to make it unclear whether we have different practices of forgiveness or a mixture of forgiveness practices and other practices of reconciliation. The ancient Greek and Chinese Buddhist traditions, and the early Jewish and Christian traditions, evolved practices of reconciliation that share important features of later practices that we call forgiveness. The contemporary practices that are commonly thought of as practices of forgiveness themselves display a wide range of variation in their features, indicating that history has carried the kaleidoscope of forgiveness forward into the present.

A parallel range of variations appears in Hindu traditions. The *Mānava Dharmaśāstra* (contemporary estimates of its dating place it around the second or third century CE) is commonly regarded as the preeminent ancient Indian text on *dharma* (righteousness, law, duty), and one of its sections states concrete maxims pertaining to acts for which one must do penance:

> When someone says "Hum" [an exclamation or a curse] to a Brahmin or addresses a superior as "you" [using a term that is too informal], he should bathe, fast the rest of the day, and placate that person by paying him obeisance. If he strikes such a man even with a blade of grass, throttles his neck with a cloth, or defeats him in an argument, he should placate him by prostrating himself on the ground.[72]

There seems here a strong resemblance to the forms of Greek reconciliation described earlier as "appeasement of anger and the relinquishing of revenge ... perceived as resting on the restoration of the dignity of the injured party, whether through compensation or gestures of deference."[73] No inner change of heart by the offender or the offended seems necessary.

On the other hand, the Sanskrit word *kṣamā*, which is most often associated with forgiveness and is alternatively translated as "forbearance" and "patience," can have much of the forward-looking connotation of a quality that enables one to do good for oneself and others. On the question of whether forgiveness is interactive or given without condition on repentance, there is a fascinating duality associated with the male and the female. Vashudha Narayan characterizes the goddess Sri (more commonly known as Lakshmi) as forgiving even when there is no repentance on the part of the soul that perpetuates the wrongdoing.[74] Though she resides as the breast jewel of Lord Vishnu and is inseparable though distinct from him, Lord Vishnu forgives only when there is repentance. Narayan points out two paradigmatic cases of forgiveness in the epic *Ramayana* (probably dating from not before 300 BCE). Sri, incarnated as Sita, the wife of Rama, forgives a crow even as it is harming her, and she also forgives the demon-women of Lanka even as they are harassing her. By contrast, Vishnu, incarnated as Rama, forgives Sugriva after he shows repentance. Sugriva offers help to Rama but later forgets his promise. Rama is angered and sends his brother Lakshmana to remind Sugriva, who then apologizes and is warmly forgiven by Rama. In the epic *Mahābhārata*, the Kaurava princes dishonor the queen Draupadi in a royal court and exult about it. The queen does not forgive them, nor is she expected to when there is no repentance. Gandhi's conception of forgiveness resonates with Sri's and with Jesus's forgiveness on the cross, the latter of which he cited in expressing his resolve to forgive the persons he anticipated would assassinate him.[75]

Normative Arguments for Different Models of Forgiveness

I question whether it is so important to clearly demarcate practices of forgiveness from others that seem to overlap in their elements, at least if the aim is to elevate one form above others. So far I have discussed the attempt to elevate one form of forgiveness as paradigmatic or "modern." There is another way to try to elevate one form, which is to argue that it is somehow morally better to practice forgiveness in that form. I remain unconvinced that there is a single morally superior form. Let me first explain why I would not be surprised by such a pluralist conclusion.

In some of my past work, I have argued that morality is that part of human culture that has an interpersonal function of promoting and regulating social cooperation in ways that do not depend on coercion or force. I have also argued that different moralities must share some general features if they are to perform their functions of coordinating human beings who tend to have characteristic sorts of motivations, but that the contingencies of culture differentiate how these general features are realized in any particular place and time. Applied to forgiveness as a response to perceived wrongdoing, my functionalist theory will lead us to expect moralities to have a variety of ways to reconcile people to each other, for us to be able to expect that some ways might be better than other ways, at least under certain conditions, but not to expect that we should find one supreme, most reasonable or justified form across the board. I have not found a good case for reserving the name *forgiveness* for any of the preferred models advocated by contemporary theorists.

Consider now the moral arguments that theorists have given for and against models of forgiveness. Griswold argues that forgiveness should be conditional on the offender taking responsibility for his wrongful acts. Failure to do so not only "adds insult to injury so far as the victim is concerned, but undermines the possibility of trusting that the offender will not turn around and repeat the injury. To forgive would then collapse into condonation."[76] Now, independently of what model of forgiveness should be adopted, I readily agree that a wrongdoer should take responsibility for his wrongful act and not add insult to injury. If Griswold's comments are to constitute an argument for making forgiveness *conditional* on such acceptance of responsibility, it must rest on the assumption that doing so will make it more likely that the offender will not turn around and repeat the injury. In other words, the assumption is that the prospect of forgiveness has sufficient motivational power to promote that outcome. Forgiveness must *matter enough* to the offender, but the kinds of contemporary cases I have cited often seem to involve offenders to whom it does not matter enough, if at all. I submit that it is very often the case that forgiveness does not matter enough to change the offender's behavior. One might simply assert that it is right for the aggrieved to maintain resentment toward the offender, regardless of whether doing so has any deterrent effect. But a blank assertion of this kind seems unpersuasive if meant as a universal claim that can be made on Confucians and Buddhists, as well as the Americans who forgave in the cases described here. If this is the best moral reason for insisting on interactive forgiveness, I submit that it is not compelling as a reason that applies across all possible applications.

If it were to be argued that interactive forgiveness is the best general paradigm for forgiveness, I submit that it is simply an unproven assumption that making one model of forgiveness the required or most desirable form of forgiveness is most conducive to fostering the kind of society one should realize. We should perhaps give people more credit for being able to apply different practices of forgiveness when appropriate. If there is an error in judgment about what form of forgiveness to apply, it lies at least as much (and, in my view, perhaps more so) in being needlessly punitive. Furthermore, when there is disagreement over who is truly the offender and who is the injured party, imposing the model of interactive forgiveness can quickly spiral into antagonism and mutual accusation that threatens the capacities of the parties to reconcile or restore their relationship.

There are morally substantive reasons in favor of unconditional forgiveness, at least on certain kinds of occasions. Sometimes it is a reason of hopefulness in the possibility that the offender is more than the terrible act he has committed and that he will find that better part of himself in the future. In addition, one might hope that the very act of extending unbidden forgiveness can itself have an influence on the offender. Jean Hampton has defended a conception of forgiveness in which "the forgiver does not wait for the wrongdoer to prove himself to be morally reborn in order to reassociate with him. Instead the forgiver *trusts* that, although he has undergone no rebirth, he is still 'good enough' despite what he has done."[77] As Hampton takes care to point out, this does not amount to condonation of wrongdoing, but a decision to see the offender, not the offense, in a new light. Whether this is a good reason depends on who is being trusted. If one trusts indiscriminately, one might destroy one's very capacity to trust, but I think that sometimes this reason for unconditional forgiveness can be a good one.

Sometimes at least part of the reason to forgive is to save oneself from corrosive anger and hatred, as suggested in the earlier discussion of the McDonald case. Anger looks backward and can prevent a person from moving forward in their life. It can be a substitute for acknowledging one's grief at what has been lost, including a loss of trust and sense of safety that can spread from fear of a single offender to one's whole social environment.[78] If one doesn't acknowledge and cope with such loss, one may be unable to escape the past.

Martha Nussbaum has presented arguments against interactive (which she calls "transactional") forgiveness of the sort that does not require repentance from the offender. Forms of interactive forgiveness influenced by the Christian tradition can be extraordinarily cruel, she points out, demanding confession, weeping, wailing, and a demonstrated sense of one's lowness and essential worthlessness. The "drama of lowness and fear has been amped up so high that there seems to be no room for personal dignity or self-respect," she writes.[79] Surely there are instances of cruelty-inflicting interactive forgiveness, and at those times it may be best for the aggrieved to try to move forward, past the wrongdoing.

However, it is hard to deny that on many other occasions interactive forgiveness can be genuinely restorative. I suspect that this might often be true of intimate personal relationships and friendships, where wrongdoing might have to be convincingly repudiated by the offender so that trust can be restored. Of course, such relationships might also constitute likely sites for exaction of the kind of cruelty Nussbaum has in mind. Withholding forgiveness in personal relationships can be a very powerful form of sustained punishment and revenge, but there seems to be no way to use the asking and giving of forgiveness as a form of restoring trust in our most meaningful relationships without being subject to its use as a weapon. And in the public sphere, where wrongdoing is continually denied and part of a history of oppression, as in many instances of police treatment of African Americans, forgiveness conditional on acceptance of wrongdoing in a collective and institutional, as well as individual, sense may be the only possibility if forgiveness is to be extended at all and social trust is to be restored.

Nussbaum casts suspicion even on unconditional forgiveness that makes no demands on the offender for repentance. Here she asks whether the anger that is forsworn in forgiveness should have been there in the first place. In her view, anger involves a kind of retributive payback wish that simply has no justification. Unconditional forgiveness can be a covert channeling of this wish: taking the moral high ground in a superior and condescending way, a way of humiliating the offender.[80] Nussbaum's questioning of the appropriateness of anger resonates with the Confucian forward-looking alternative for the victim of wrongdoing: to focus on what is the appropriate response from the self rather than on what the offender has done to the self.[81] But we saw earlier that this kind of response might be consistent with a kind of unconditional forgiveness born of a desire to move forward with one's life and to address in a forward-looking manner the kind of wrongdoing that has caused one's injury. In fact, Nussbaum grants that the unconditional forgiveness conferred by the family of a victim in the South Carolina shootings did not show any hint of a payback wish, so it is only sometimes that unconditional forgiveness takes perversely vengeful form. I wonder if *any* practice of reconciling with wrongdoers is free from misuse. After all, to claim to have done away with any need for anger, and therefore any need for forgiveness, might be a way of appearing to take the *highest* moral ground but in the *most* superior and condescending way.

I agree with Owen Flanagan that we foolishly give far too much cultural permission to becoming angry,[82] and if anything the recent and ongoing state of politics in US society underscores his point. But the cultural permission given to anger seems to vary with gender and race. It is far too often less acceptable for women and girls to express anger. Ann Kring, in a survey of the psychological literature on gender and anger, reports that men and women self-report comparable degrees of anger but that women experience more shame and embarrassment after experiencing anger.[83] The Me Too movement, in which increasing numbers of women from all walks of life have come to reveal instances of sexual aggression against them, demonstrates how powerful the norm against expressing anger can be for women.[84] The first few declarations, most of them by women with public visibility and credibility, opened a floodgate of further declarations. Perhaps a new kind of permission was given, to turn around the direction of blame from the self, for letting something bad happen, to the perpetrator. In these instances, anger and resentment not only seem intelligible and perfectly appropriate but perhaps also *necessary* to overcome fear of retaliation by powerful perpetrators or of having one's vulnerability exposed.

Of course, resentful anger can easily get out of hand, so even if it serves a good purpose and enables one to overcome inhibitors to right action, it must be controlled. Gandhi wrote that he had "learnt through bitter experience the one supreme lesson to conserve [his] anger, and as heat conserved is transmuted into energy, even so our anger controlled can be transmuted into a power which can move the world."[85] He said this in support of the policy of "non-cooperation" with the oppressive British rule of India as an alternative to violent resistance. He associated the desired transmutation of anger with discipline and self-sacrifice, and contrasted it with the "mere expression of angry feeling."[86] Gandhi's grandson, Arun Gandhi, tells how Gandhi taught him that anger is good, like gas to the automobile, helping one to move forward and get to a better place. Without it, he said, one would not be motivated to rise to a challenge or to define what is just and unjust.[87]

Gandhi instructed his grandson to keep a journal, and for each time he felt great anger, he should write down who or what caused those feelings and why he reacted so angrily. The goal was to get to the root of the anger, which was necessary to find the solution. Further, in understanding the source of conflict, one has to acknowledge the point of view of everyone involved. This wasn't giving in, Gandhi emphasized, but instead trying to find a solution that didn't lead to more anger and resentment. Attacks, criticism, and threats of punishment backfire with both children and adults. Arun writes that Gandhi told him that "being able to understand another's viewpoint and forgive is the sign of real strength."[88] Being able to reflect in this way, of course, did not come easily, so Gandhi taught his grandson to sit in a quiet room without distractions and to concentrate on something lovely in front of him. Could he close his eyes and keep the image in his mind? Gandhi instructed him to practice at holding it longer and longer. It is interesting that the two things Gandhi taught his grandson correspond to findings about the effects of naming one's emotions and meditating mindfully on one's present feelings, thoughts, and sensations. Both have the effect of increasing one's ability to regulate their emotions. Both have the effects of activating the prefrontal cortex and dampening activity in the amygdala, the area of the brain most responsible for the impulsive reactiveness of anger.[89]

Arun Gandhi gives an example of what his grandfather's teaching later enabled him to do. As a young man, he was living in South Africa. When he went to India to visit relatives, he met and fell in love with the woman who would become his wife. But because of apartheid, she was not allowed to go back with him to South Africa, even though he was a citizen of that country. He made the difficult choice of staying with his wife in India and not returning to his widowed mother and his sisters in South Africa. Some years later, he met Jackie Basson, a member of the South African parliament, who asked Gandhi to show him the city of Mumbai. Gandhi's impulse was to throw the man overboard from the ship he had arrived on, but he controlled his anger and helped Basson and his wife have a very nice tour. At the end, Gandhi explained his grievance against the South African government to the Bassons and how it had pulled his family apart. Jackie Basson responded by speaking out against apartheid when he got back to South Africa. Though the ruling party disowned him and he lost the next election, he remained steadfast in his opposition to apartheid.[90]

I am unable to condemn all forms of forgiveness because I am reluctant to advocate eliminating all anger, and in particular the form of anger that is related to resentment. I wonder whether we can do without the propulsive energy it provides, though I think we often do far too little in the way of trying to understand and teach how to control it in the way Zhu Xi and Gandhi advocated. It is true that such a task would be formidable, but perhaps less so than the task of eliminating anger. For the reasons given earlier in this chapter, I am reluctant to condemn unconditional forgiveness in favor of interactive forgiveness, as some have argued. Yet, as I have also argued, there are times when interactive forgiveness seems called for.

After the genocide by Hutus against Tutsis, in which nearly a million people died, Rwanda launched a continuing effort toward reconciliation in which members of both groups were counseled after many months. A perpetrator would make a formal request for forgiveness, and if accepted by the survivor, the perpetrator and his family and friends would typically bring a basket of food and drink, and there would be song and dance. Various reasons for granting forgiveness were given by survivors, and they invoke all the forms of forgiveness-related phenomena described earlier.

Karorero, a survivor, forgave Sinzikiramuka, who was complicit in the killing of Karorero's brother, because, as he put it, "When someone is full of anger, he can lose his mind. But when I granted forgiveness, I felt my mind at rest." Mukanyandwi used to hate Mudaheranwa, who burned her house and tried to kill her and her children. But when he came to her, knelt down, and asked for forgiveness, she was moved by his sincerity. "Now, if I cry for help, he comes to rescue me. When I face any issue, I call him." Kampundu forgave Nzabamwita, who damaged and looted her property and whose father participated in the killing of her children, for a starker reason. She stated, "I realized that I would never get back the beloved ones I had lost. I could not live a lonely life—I wondered, if I was ill, who was going to stay by my bedside, and if I was in trouble and cried for help, who was going to rescue me? I preferred to grant pardon." Makabutera forgave Habyarimana, who was involved in the killing of her children, and explained, "It took time, but in the end we realized that we are all Rwandans. The genocide was due to bad governance that set neighbors, brothers and sisters against one another. Now you accept and you forgive. The person you have forgiven becomes a good neighbor. One feels peaceful and thinks well of the future." Mukamusoni forgave Ntambara, who killed her child, saying,

> He did not do it by himself—he was haunted by the devil. I was pleased by the way he testified to the crime instead of keeping it in hiding, because it hurts if someone keeps hiding a crime he committed against you. Before, when I had not yet granted him pardon, he could not come close to me. I treated him like my enemy. But now, I would rather treat him like my own child.[91]

That people spread the word *forgiveness* over a variety of different but related practices of reconciliation feels like a force of nature that philosophers can do little to control. We can ask what reasons there are to engage in these practices, whatever we call them, and on this matter I find it hopeless and ill-advised to champion one model only. If various reasons are given and correspond to different forms of forgiveness, I can see no case for forbidding a varied repertoire of different practices of forgiveness and to accept the task of more particularized judgment as to when and for whom a given practice is most warranted. Individuals, in their political, social, cultural, and personal circumstances, might find any one of these forms of reconciliation appropriate and justifiable. It might be objected that there might still be a single form or small set of them that are usually most advisable, but as the foregoing might suggest, I see no good case for such a conclusion. Owen Flanagan has drawn attention to the advantages of "superwide" cross-cultural reflective equilibrium (which is essentially comparative philosophy) that helps us to avoid parochial justification of our ways of doing things.[92] Tapping into forgiveness-related practices across cultures not only can give us an idea of alternative ways of doing things but can also make us realize that our own ways are not as univocal as we thought they were. I find it good to have a repertoire of practices from which we can draw in different circumstances. Not only is forgiveness elusive, it is good that it is so.

Notes

1. Flanagan, *The Geography of Morals*, Kindle locations 3949–3952.

2. Ibid., Kindle locations 372–375.

3. My thanks go to Wenqing Zhao, Tom Polger, and Ben Sarbey for very helpful comments on previous drafts of this chapter. Failures to be persuasive are, of course, mine alone.

4. Strawson, *Freedom and Resentment and Other Essays*, Kindle locations 672–675, 849.

5. Flanagan, *The Geography of Morals*.

6. Ibid., Kindle location 3321.

7. Strawson, *Freedom and Resentment and Other Essays*, 6.

8. Ibid., 15.

9. Konstan, *Before Forgiveness*, 11.

10. Ibid., 59.

11. Ibid., 14.

12. Ibid., 26; Aristotle, *Nicomachean Ethics*, 1109b18–1111a2.

13. Konstan, *Before Forgiveness*, 29.

14. Ibid., 32.

15. Griswold, *Forgiveness*, 7–8.

16. Shun, Resentment and forgiveness in Confucian thought, 28.

17. Ibid., 23.

18. Ibid., 24.

19. Xunzi, *Sibubeiyao* edition, 12.12b.

20. Shun, Resentment and forgiveness in Confucian thought, 30.

21. Ibid.

22. See Eno, *The Analects of Confucius*, 79.

23. See Slingerland, *Confucius*, 167–168, following third-century commentator He Yan.

24. Huang, *Confucius*, 37.

25. Slingerland, *Confucius*, 167.

26. Huang, *Confucius*, 37.

27. See Slingerland, *Confucius*, 167, and Huang, *Confucius*, 37–45, for an extensive discussion of the different interpretations given to this passage.

28. Huang, *Confucius*, 42. Huang reinforces this reading with citations of *Mengzi* 3A1 and 3A4 and the *Zuozhuan* Duke Xiang Year 7 with a similar association between uprightness and making others so; see Huang, *Confucius*, 41–42, and *Mengzi* 3A1, 3A4.

29. I thank Xin Zhang for bringing up this passage in *Mengzi* with me.

30. Kim calls the resentment in question "affective resentment" to distinguish it from self-focused forms. He notes that the context that makes affective resentment intelligible is a conception of the family not as a sentimentalized arena of static enlarged affection but rather as a "dynamic moral arena in which each person experiences personal moral growth through dialogic interactions"; see Kim, *Confucian Democracy in East Asia*, 67.

31. Griswold, *Forgiveness*, 130. See also Shun, Resentment and forgiveness in Confucian thought, 32.

32. Goldstein, Steven McDonald, paralyzed officer who championed forgiveness, dies at 59, A21.

33. Ibid.

34. Griswold, *Forgiveness*, 121.

35. Ibid., xvi.

36. There is one other important possibility for privileging interactive forgiveness that I will discuss toward the end of this chapter, which is that there is a moral case for favoring that form of forgiveness.

37. In personal correspondence, Shun has indicated to me agreement that forgiveness might take forms other than the one that formed the basis of his contrast with the Confucian stance.

38. Ziporyn, *Zhuangzi*, 35.

39. Rye et al., Religious perspectives on forgiveness, 21–22.

40. Ibid., 27.

41. Cited in Rye et al., Religious perspectives on forgiveness, 27–28; Buddhadatta, *Dhammapada*, 2.

42. Thanks to Don Davis for informing me of Hallissey's thinking on forgiveness and Buddhism.

43. Tsongkhapa quoting from Asanga, in Wayman, *Ethics of Tibet*, 143. Thanks to Jens Schlieter for providing me with this reference.

44. Khoroche, *When the Buddha Was a Monkey*, 177, cited in Rye et al., Religious perspectives on forgiveness, 22, 39.

45. Khoroche, *When the Buddha Was a Monkey*, 202–203, cited in Rye et al., Religious perspectives on forgiveness, 34, 40.

46. Konstan, *Before Forgiveness*, 91.

47. Ibid., 91–94.

48. Kohler, Atonement, cited in Konstan, *Before Forgiveness*, 105.

49. Konstan, *Before Forgiveness*, 113.

50. The Mishnah, a second-century Palestinian law code, as cited in Newman, The quality of mercy, 159.

51. Maimonides, *The Code of Maimonides (Misneh Torah)*, book 4, *Laws of Repentence*, 2:4, 9–11; cited in Newman, The quality of mercy, 160.

52. Newman, The quality of mercy, 167.

53. Konstan, *Before Forgiveness*, 116.

54. Luke 23:34, cited in Konstan, *Before Forgiveness*, 120.

55. Ibid., 123.

56. Ibid., 112.

57. Ibid.

58. Ibid., 113.

59. Ibid., 118.

60. Ibid., 122.

61. Ibid.

62. Ibid., 119.

63. Berman, "I forgive you."

64. Bråkenhielm, *Forgiveness*, 16, cited in Konstan, *Before Forgiveness*, 121.

65. Konstan, *Before Forgiveness*, 121.

66. Mindock, Heather Heyer.

67. Konstan, *Before Forgiveness*, 147.

68. Ibid., 155.

69. Ibid., 157.

70. Sermon IX from Butler, *The Works of Bishop Butler*, 100.

71. Shun, Resentment and forgiveness in Confucian thought.

72. *Yājñavalkya Smṛti*, 3.291. See Olivelle, *Manu's Code of Law*, 226, 346. Thanks to Patrick Olivelle for offering his view on the presence of forgiveness in Indian traditions and for providing me with the reference to the passage in Manu.

73. Konstan, *Before Forgiveness*, 59.

74. Narayan et al., Religious perspectives on forgiveness, 34–35.

75. R. Gandhi, *Mohandas*, Kindle locations 13588–13589.

76. Griswold, *Forgiveness*, 49.

77. Murphy and Hampton, *Forgiveness and Mercy*, 84.

78. On forgiveness as addressing breaches of trust, see especially Walker, *Moral Repair*.

79. Nussbaum, *Anger and Forgiveness*, 73.

80. Ibid., 76–77.

81. Nussbaum appears to take the more radical stance that anger simply is bad, but she has written that if there is a kind of "transitional" anger that is ameliorative and forward-looking, it might even be good. See Nussbaum, Transitional anger; Nussbaum, *Anger and Forgiveness*.

82. Flanagan, *The Geography of Morals*, Kindle location 3296.

83. Kring, Gender and anger.

84. Harassed; Farrow, From aggressive overtures to sexual assault.

85. M. Gandhi, Speech on non-co-operation resolution, 252.

86. Ibid., 252–253.

87. A. Gandhi, *The Gift of Anger*, 17–18.

88. Ibid., 21–22.

89. Creswell et al., Neural correlates of dispositional mindfulness during affect labeling; Lieberman et al., Putting feelings into words; Lieberman et al., Subjective responses to emotional stimuli during labeling, reappraisal, and distraction.

90. A. Gandhi, *The Gift of Anger*, 29–30.

91. All quotations are from Dominus, Portraits of reconciliation.

92. Flanagan, *The Geography of Morals*, Kindle location 7104.

Bibliography

Analects. Chinese Text Project, https://ctext.org/analects.

Berman, Mark. "I forgive you": Relatives of Charleston church shooting victims address Dylann Roof. *Washington Post*, June 19, 2015. https://www.washingtonpost.com/news/post-nation/wp/2015/06/19/i-forgive-you-relatives-of-charleston-church-victims-address-dylann-roof/?utm_term=.14893f9d9495.

Bråkenhielm, Carl Heinz. *Forgiveness*. Translated by Thor Hall. Minneapolis, MN: Fortress Press, 1993.

Buddhadatta, Ambalangoda Polwatte, trans. *Dhammapada: An Anthology of Sayings of the Budda*. Colombo: Apothecaries, n.d.

Butler, Bishop [Joseph]. *The Works of Bishop Butler*. Edited by David E. White. Rochester, NY: University of Rochester Press, 2006.

Creswell, J. David, Way, Baldwin M., Eisenberger, Naomi I., and Lieberman, Matthew D. Neural correlates of dispositional mindfulness during affect labeling. *Psychosomatic Medicine* 69 (2007): 560–565.

Dominus, Susan. Portraits of reconciliation. *New York Times Magazine*, April 4, 2014. https://www.nytimes.com/interactive/2014/04/06/magazine/06-pieter-hugo-rwanda-portraits.html?r=0.

Eno, Robert. *The Analects of Confucius. An Online Teaching Translation*. 2015. https://scholarworks.iu.edu/dspace/bitstream/handle/2022/23420/Analects_of_Confucius_%28Eno-2015%29-updated.pdf?sequence=2&isAllowed=y.

Farrow, Ronan. From aggressive overtures to sexual assault: Harvey Weinstein's accusers tell their stories. *New Yorker*, October 23, 2017. https://www.newyorker.com/news/news-desk/from-aggressive-overtures-to-sexual-assault-harvey-weinsteins-accusers-tell-their-stories.

Flanagan, Owen. *The Geography of Morals: Varieties of Moral Possibility*. Kindle edition. Oxford: Oxford University Press, 2017.

Gandhi, Arun. *The Gift of Anger: And Other Lessons from My Grandfather Mahatma Gandhi*. New York: Jeter Publishing, 2017.

Gandhi, Mahatma. Speech on non-co-operation resolution, Calcutta Congress, September 8, 1920. In *Collected Works of Mahatma Gandhi*, vol. 21. New Delhi: Publications Division, Government of India, 1999. http://www.gandhiashramsevagram.org/gandhi-literature/mahatma-gandhi-collected-works-volume-21.pdf.

Gandhi, Rajmohan. *Mohandas: True Story of a Man, His People*. Kindle edition. New York: Penguin, 2007.

Goldstein, Richard. Steven McDonald, paralyzed officer who championed forgiveness, dies at 59. *New York Times*, January 11, 2017, A21.

Griswold, Charles. *Forgiveness: A Philosophical Exploration*. New York: Cambridge University Press, 2007.

Harassed. *New York Times*, 2018. https://www.nytimes.com/spotlight/harass.

Huang, Yong. *Confucius: A Guide for the Perplexed*. London: Bloomsbury Academic, 2013.

Khoroche, Peter, trans. *When the Buddha Was a Monkey: Arya Sura's Jatakamala*. Chicago: University of Chicago Press, 1989.

Kim, Sungmoon. *Confucian Democracy in East Asia: Theory and Practice*. Cambridge: Cambridge University Press, 2014.

Kohler, Kaufmann. Atonement. In *The Jewish Encyclopedia, 1901–1906*, vol. 1, 275–284. http://www.jewishencyclopedia.com/view.jsp?artid=2092&letter=A&Search=atonement.

Konstan, David. *Before Forgiveness: The Origins of a Moral Idea*. New York: Cambridge University Press, 2010.

Kring, Ann M. Gender and anger. In *Gender and Emotion: Social Psychological Perspectives*, edited by Agnetta H. Fischer, 211–231. Studies in Emotion and Social Interaction. Cambridge: Cambridge University Press, 2000.

Lieberman, Matthew D., Eisenberger, Naomi I., Crockett, Molly J., Tom, Sabrina M., Pfeifer, Jennifer H., and Way, Baldwin M. Putting feelings into words: Affect labeling disrupts amygdala activity to affective stimuli. *Psychological Science* 18 (2007): 421–428.

Lieberman, Matthew D., Inagaki, Tristen K., Tabibnia, Golnaz, and Crockett, Molly J. Subjective responses to emotional stimuli during labeling, reappraisal, and distraction. *Emotion* 11, no. 3 (June 2011): 468–480. https://doi.org/10.1037/a0023503.

Maimonides, Moses. *The Code of Maimonides (Misneh Torah)*. Book 4, *Laws of Repentance*, translated by Hyman Klein. New Haven, CT: Yale University Press, 1954.

Mengzi. Chinese Text Project, https://ctext.org/mengzi.

Mindock, Clark. Heather Heyer: Father of Charlottesville victim says he forgives the white supremacist who killed his daughter. *The Independent*, August 15, 2017. http://www.independent.co.uk/news/world/americas/charlottesville-protests-latest-heather-heyer-father-mark-forgives-james-fields-killing-his-daughter-a7894996.html.

Murphy, Jeffrie G., and Hampton, Jean. *Forgiveness and Mercy*. Cambridge: Cambridge University Press, 1984.

Newman, Louis E. The quality of mercy: The duty to forgive in the Judaic tradition. *Journal of Religious Ethics* 15, no. 2 (1987): 155–172.

Nussbaum, Martha. *Anger and Forgiveness: Resentment, Generosity, Justice*. New York: Oxford University Press, 2016.

Nussbaum, Martha. Transitional anger. *Journal of the American Philosophical Association* 1, no. 1 (2015): 41–56.

Olivelle, Patrick. *Manu's Code of Law: A Critical Edition and Translation of the Mānava Dharmaśāstra*. Oxford: Oxford University Press, 2005.

Paralyzed officer forgives teen-age assailant. *New York Times*, March 2, 1987. http://www.nytimes.com/1987/03/02/nyregion/paralyzed-officer-forgives-teen-age-assailant.html.

Rye, Mark S., Pargament, Kenneth I., Ali, M. Amir, Beck, Gut L., Dorff, Elliot N., Hallisey, Charles, Narayan, Sudha, and Williams, James G. Religious perspectives on forgiveness. In *Forgiveness: Theory, Research, and Practice*, edited by Michael E. McCullough, Kenneth I. Pargament, and Carl E. Thoresen, 17–40. New York: Guilford Press, 2000.

Shun, Kwong-loi. On anger: An essay in Confucian moral psychology. In *Returning to Zhu Xi: Emerging Patterns within the Supreme Polarity*, edited by David Jones and Jinli He, 299–324. Albany: State University of New York Press, 2015.

Shun, Kwong-loi. Resentment and forgiveness in Confucian thought. *Journal of East-West Thought* 4, no. 4 (2014): 13–35.

Slingerland, Edward. *Confucius: The Essential Analects—Selected Passages with Traditional Commentary*. Indianapolis, IN: Hackett, 2006.

Strawson, P. F. *Freedom and Resentment and Other Essays*. New York: Routledge, 2008.

Walker, Margaret Urban. *Moral Repair: Reconstructing Moral Relations after Wrongdoing*. New York: Cambridge University Press, 2006.

Wayman, Alex. *Ethics of Tibet: Bodhisattva Section of Tsong-Kha-Pa's Lam Rim Chen Mo*. State University of New York Series in Buddhist Studies. Albany: State University of New York Press, 1991.

Xunzi. *Sibubeiyao* edition. Taipei: Zhonghua shuju, 1965.

Ziporyn, Brook, trans. *Zhuangzi: The Essential Writings, with Selections from Traditional Commentaries*. Indianapolis, IN: Hackett, 2009.

9 Wagging Tails and Riding Elephants: Why Study Non-Western Philosophy?

Philip J. Ivanhoe

> The development of consciousness in human beings is inseparably connected with the use of metaphor. Metaphors are not merely peripheral decoration or even useful models, they are fundamental forms of our awareness of our condition.... It seems to me impossible to discuss certain kinds of concepts without resort to metaphor, since the concepts are themselves deeply metaphorical and cannot be analyzed into non-metaphorical components without a loss of substance.
>
> —Iris Murdoch, *The Sovereignty of Good*

Introduction

I want to advance two arguments connected, though in different ways, with the value of studying non-Western philosophy.[1] If the arguments are sound, they show not only that it can be helpful to study non-Western philosophy but also that, at least for ethical and political theory, it would be irresponsible not to do so. I believe both these claims are true.

My first argument is that those who advocate a *wide* "reflective equilibrium"[2] approach to ethics and political theory, those who seek to begin philosophical reflection by sympathetically understanding and critically evaluating the best views currently in play, are not being true to their own principles or are assuming, without evidence or argument, that there are no wise views outside the Western canon. I take it as evident that the latter assumption is both irresponsible and absurd. Now, some will respond that this is not the way to approach ethical or political theory, and we can debate the merits pro and con of such an approach, but it is simply a fact that many well-known philosophers and political thinkers today defend and practice this kind of approach—philosophy and political science departments rely on classes that consist of surveys of works from the Western canon, which strongly implies that this is how one begins the study of these disciplines—so the criticism applies at least to a very large group of, and arguably almost all, contemporary philosophers and political theorists.[3] The argument can and should be extended to include contemporary thinkers who argue that ethical and political theory should be based on, or at least attentive to, work in the social sciences. In this case, the thought is that such approaches take among their primary objects of study the values and norms that people actually have and practice, and this leads one to at least begin with the best views currently available and in play. Among philosophers who employ a version of wide reflective equilibrium—which comprehensively embraces not only philosophy but the social sciences, empirical and evolutionary psychology, and cognitive neuroscience—are people like Owen Flanagan and, as I will show, he and his work are particularly revealing for exploring the connection to non-Western philosophy.

My second argument concerns two metaphors Jonathan Haidt offers to represent the relationship between reason and emotion in ethical justification and what these imply, conceal, and preclude. (This is where the quotation from Iris Murdoch that serves as the epigraph of this chapter comes in.) Haidt has become well known and in some quarters notorious for arguing that moral judgment and justification are primarily and almost wholly matters of "fast" emotionally based responses and have very little to do with "slow" rational deliberation or decision.[4] The latter purportedly are almost always post facto and function as a kind of rearguard action to defend decisions that our fast-acting emotional system already has made. This of course would pose significant problems for just about every moral and political theory currently defended by contemporary thinkers, which unsurprisingly has drawn criticism from some quarters.[5] Among such critics are Owen Flanagan, who argues for a more robust and important role for reason.[6] What I have to say about Haidt's views supports Flanagan's position, which I share, but this is not my primary concern. My core interest is to note that Haidt and those who take part in the widening debate his views have generated in one way or another maintain and remain stuck in the venerable dichotomy between reason and emotion. I will show how Haidt's choice of metaphors both reflects and in certain respects precludes or obscures this important point, which leads him as well as his defenders and critics to fail to see how reasoning almost always enters into and informs, through culturally situated practice and tradition, our emotional and ethical lives. I then will show how serious engagement with East Asian philosophical traditions offers a way out of this debilitating dichotomy and toward philosophical and psychological views that are more plausible.[7]

Moral Modularity

Flanagan's interest in moral modularity is long-standing and is clearly stated in his classic work *Varieties of Moral Personality: Ethics and Psychological Realism*.[8] Initially inspired by the work of Chomsky and Fodor,[9] Flanagan took up and developed the idea that human beings are endowed with domain-specific, modular moral processors. These processors are primarily brain-based systems that are to a significant extent functionally autonomous (i.e., dedicated to certain phenomena that they track and respond to largely without the aid of other parts of the self and at least initially fairly well encapsulated); that is to say, not readily susceptible to control or direction from other modules or some overall executive agency. While softening the strong isolated silo description of modularity characteristic of Chomsky's and Fodor's work, which is attributable at least in part to the particular human capacities that were the focus of their investigations, and allowing for greater permeability and coordination among moral modules, Flanagan saw clearly and showed precisely how revolutionary such a picture might be for traditional approaches to ethics by asking, "The obvious question, however, is whether moral competence is appropriately construed as a coherent domain in its own right or whether it is, in reality, just a convenient term which depicts a multifarious set of competencies, each possibly with its own learning story."[10]

The "unitary-competence" view being challenged here is characteristic of the Stoics and most Enlightenment ethicists as well as the theories of psychologists such as Piaget and Kohlberg, whose research was widely regarded as providing empirical support for such views. The ethical theories of neo-Aristotelians as well as pragmatists like Dewey are consistent with versions of the alternative "multiple-competence" view. Flanagan offers a variety of reasons for thinking that some version of the multiple-competence view is more plausible (e.g., it is difficult to identify any essential feature of moral problems, character, or beliefs)—morality and character "are more what psychologists call cluster concepts than they are classical natural kind terms";[11] moral traits are acquired and function in very different ways; there is considerable "gappiness" in actual moral character—some people regularly show great compassion but appear unmusical or even deaf when it comes to justice or courage. This last point, along with others, shows that the "multiple-competence" view does not provide strong support for many traditional forms of virtue ethics either, for, in addition to tending to rely on a single strict and inflexible teleology, such views also tend to see much greater consistency and range in individual virtues and much greater unity and relationship among the virtues than the empirical evidence warrants.[12] A "multiple-competence" modularity theory that allows for good as well as bad natural tendencies, though, can offer a plausible psychological basis for both virtues and vices. In later work, Flanagan also brings into play ideas and evidence articulated by contemporary psychologists such as Jonathan Haidt, whose work we will discuss in more detail, about the existence of specific sets of innate moral modules that can be tuned up or down within different moral traditions, which also militate against most traditional forms of virtue ethics.[13]

There are fascinating and deeply challenging questions and puzzles surrounding the general claims of the Moral Modularity Hypothesis (MMH). Some of these, such as the degree of binding that exists among different innate modules, apply equally and fundamentally to nonmoral abilities such as vision, while others, such as whether and in what sense a given module needs to be engaged or employed in order to produce a viable ethical system or the ways in which culture and history contribute to or shape such decisions, seem to apply at least more directly and dramatically to moral modules than to nonmoral ones.[14] Evolutionary theory has a great deal to contribute to our understanding of modularity in all its forms, but as is often the case, evolution contributes but is by no means decisive, for, as Flanagan and others have pointed out, while evolution surely has bequeathed to us the capacities with which we begin life, it does not control either the circumstances within which they operate or the ways in which we develop and deploy our capacities.[15] Innate modules for vision or for detecting and responding to suffering developed over extensive periods of time and enabled our ancestors to survive and develop, but we exercise these inherited capacities in profoundly different circumstances. Our innate tendencies to gorge on sugar and animal fat worked very well when these substances were extremely difficult to find, but such tendencies strongly incline us to consume unhealthy amounts of them in today's environment. Moreover, we develop and deploy our innate modules very differently and with new and previously unforeseen goals in mind. Color vision served important functions in the distant past, helping us to identify ripe fruit or rancid meat, but today we use it to know when to brake or continue on through an intersection. We also cultivate and deploy it in the production and appreciation of a remarkably broad variety of visual arts that motivate and satisfy us in a range of profound ways that have nothing at all to do with the original value of color vision. Such differences between our ancestral past and current life show that the evolutionary goals of survival and reproduction offer only the most basic, open-ended, and often defeasible standards for deciding how modern human beings can or should live. The older, more basic conception of fitness no longer describes what we choose to do; evolution has bequeathed to us capacities that equip and compel us to look beyond survival, to seek, discover, and invent more robust and open-ended forms of the good life in environments we have profoundly altered, shaped, and manufactured.

The MMH, as described by Flanagan, leaves us with many questions, but it offers a number of important insights and answers to ponder, appreciate, and apply. According to this view, morality is much more a matter of drawing on and developing a set of innate human tendencies, something very much like what P. F. Strawson called "reactive attitudes,"[16] and, through an ongoing process of reflective experience identifying how to adjust these up or down in strength, express them in various forms and bring them together in different permutations and combinations to address the kinds of challenges creatures like us tend to face in ways that yield viable, satisfying, and enjoyable forms of life. Now admittedly there are quite a number of parameters in play here that can be satisfied in mutually incompatible ways, but that does not entail any strong form of relativity, only a natural, healthy pluralism.[17] There is no single, fixed course of development that leads inexorably from the nascent "sprouts" of our innate moral repertoire to an ideal form of human life. It is not altogether clear that every moral module needs to be developed and deployed;[18] it is perfectly consistent with the hypothesis that informed and reflective human beings will come to judge that some sprouts no longer contribute in the ways or to the extent that they did successfully in the past. Some moral sprouts might even come to be viewed as moral weeds.[19] This and other features of Flanagan's conception of MMH lead us to explore the connection between his moral theory and comparative philosophy.

As noted, the MMH leaves us without any algorithm for deciding which moral inclinations we should develop, to what extent we should develop them, or how to relate them to one another. As many have noted, we lack a clear natural teleology in regard to morality; in this sense, we cannot move from how things are to how we should be, from is to ought, without, as Hume noted, offering some explanation of how to make this passage. Nevertheless, like the theory of natural selection itself, the MMH does not do away with the notion of teleology altogether, which is one reason it does not support strong relativism. As a number of people have pointed out,[20] Darwin still saw himself as pursuing a teleological account of the evolution of living organisms. Within a relatively stable environment, traits that enable an organism to survive and reproduce more successfully are selected for the functions they provide. Moreover, because we talk about this process in terms of the relative *advantage* of such traits, values are clearly in play. The problem with applying such an analysis to the case of morality lies in the fact, noted earlier and clearly discussed by Flanagan in terms of the difference between "fitness" and "flourishing,"[21] that human beings are not satisfied, happy, or content with lives that aim only at survival and reproduction.[22] It's not that we don't value such things; that is not the point. Rather, the point is that they are not the only or trumping values that human beings hold dear and pursue. This is not solely or even primarily a hortatory claim; it is an empirical claim. This, in fact—and it is a contingent fact—is what distinguishes us from every other creature on earth; we are "evaluative animals."[23] In light of the range and intensity of the things we value, support, advocate, sing praises to, and pursue, it is not an exaggeration to say we are promiscuously valuing creatures.

Given that human beings, by nature, tend to value morality in its various guises, we are justified in claiming that the search for and pursuit of moral ends describe a clear though general and still evolving natural teleology.[24] This requires further explanation. First, we must keep in mind that "moral ends" here refers not to a single rule or principle or set of rules and principles derived prior to human experience (like the rules and principles of traditional deontology or consequentialism) but a collection or cluster of goods that are widely seen as constitutive of individual and collective well-being. Second, we must keep in mind that how we tune, arrange, and instantiate our innate moral tendencies depends in large measure on the accumulated and ongoing reflective experiences of people in particular historical cultures and traditions. In light of these factors, the proper work of ethics and political theory is not primarily much less exclusively abstract theoretical speculation, though that has its place, but systematic reflection, analysis, critique, and development of a fundamentally empirically based inquiry drawing on biology, neuroscience, psychology, religion, anthropology, literature, and other fields aimed at discovering the bases and forms of moral values. As a general approach, this should sound familiar, for it basically describes the method of *saving the phenomena* first described and employed by Aristotle, which is the progenitor of the wide reflective equilibrium I advocate. We are not to passively register and accept what we find in either our first or second nature; both are open to and invite critical evaluation and creative reformulation and extension. Nevertheless, these are the starting points for ethical and political reflection. This is a good way to describe how Flanagan proceeds in all his work, and it shows how his sustained and productive interest in non-Western philosophy, originally focused mostly on Buddhism but now also including Chinese Mohist and Legalist philosophy as well as Chinese and Korean Confucianism, follows naturally from such an approach.[25]

The relationship between saving the phenomena and comparative philosophy is obvious and is something we will return to at the conclusion of this chapter. Here I simply want to note that it is nothing short of astounding that so few seem to recognize that the careful and sympathetic consideration of other traditions is an obvious imperative for anyone who advocates this more naturalized method in ethics and political theory. This kind of imperative to explore other cultures and traditions, both historical and contemporary, is well understood and widely practiced in fields such as psychology.[26] My point here is that there is a clear reason why this is the case. In order to successfully identify and describe general psychological features of human beings, psychologists need broad and varied samples of human behavior under a range of conditions. In recent years, most psychologists have come to appreciate just how narrow, parochial, and idiosyncratic their samples often are.[27] It is time for philosophers also to recognize how narrow, parochial, and idiosyncratic their "intuitions" are about core issues in ethics and political theory.[28] Human beings have been working independently on many of the same problems in ethics and politics at different times, places, and circumstances throughout history. "Wise observers of the human condition"[29] clearly saw truths about us that still apply and are valuable to us today, and different cultures and traditions within them have invented, refined, and carried on individual practices and more general styles of human living that offer distinctive opportunities for human satisfaction, edification, and joy.[30] Participating in the creation, improvement, and continuation of various traditions is itself a *new* and profound source for human flourishing, and we must look for and will find examples of such values throughout the world. There are many very different, highly successful societies around the world whose members are well acquainted with Western values and civilizations but prefer their own distinctive norms, values, and practices. Why would any open-minded person committed to exploring ethical and political values *in principle* constrain their exploration and consideration of sources only to their home tradition? Theological geniuses such as Augustine, Aquinas, and Leibniz[31] placed no such constraints on their thought, so why do so many contemporary, purportedly more objective, scientific, and open-minded philosophers do so? It makes one wonder to what degree the provincialism of much of contemporary ethical and political theory is more a manifestation of the lack of education, experience, empathy, and imagination than just a failing of theoretical insight (though it surely is both).[32]

One might respond to the challenge I have raised by declaring that one simply is not committed to an exploration in principle of other sources of value and instead is only interested in saving the phenomenon in the case of Western ethical and political values.[33] If one accepts the idea that there is indeed a "healthy relativism" in regard to such values, then it is not immediately clear that there is a need to study other cultures. As long as one is able to save the phenomenon and justify one's home ethical and political system, one might think that there is no real work left to be done. Of course, there are costs to holding such a view, and among them is that one must abandon any pretension to universal prescription and any expectation that those outside one's particular tradition should find one's views appealing or even acceptable.[34] I do not, however, see how those making such an appeal could have adequate grounds for believing that they have a good grasp of what the available options in fact are (i.e., that they have properly identified what basic moral modules are available, much less how they might be deployed or tuned). Without such knowledge, one could not be confident that one's particular version of saving the phenomenon is as good as other available options and thus defensible in its own right. For example, imagine a society in which compassion reigns supreme (highly tuning up something like a suffering/care module) and no attention whatsoever is paid to our capacity and tendency to seek justice (based on our fairness/reciprocity module).[35] Without engaging in comparison, could a person in such a society plausibly claim to understand what kinds of good lives are available? In order to be confident that one's particular version of saving the phenomenon is at least as good as others, it seems that one must in fact engage in comparative philosophy. The more insular position I have described is only something one might appeal to at the end of a process of comparative study, though there are good reasons to think that the process of such study itself will preclude at least certain forms of such an appeal.[36]

Turning to the particular case of Flanagan and his work, it is important for people on both sides of the East-West divide to understand and appreciate the nature and seriousness of his various forays into non-Western philosophy. These are not mere fancies or diversions but rather an integral component of the naturalizing turn and the methodology of saving the phenomena. Those on the Eastern side need to see that his interests in non-Western philosophy manifest not a desire to debunk (though on certain issues he does not and cannot shy away from raising strong objections) but a desire to learn from and not just about other conceptions, approaches, and traditions of human value. One clear piece of evidence in support of this understanding of his work is that he applies the same rigorous analysis to his own home traditions. If something like MMH is the best way to think about the nature and practice of morality and how to engage in systematic critical moral reflection and theory building, we must "go empirical" in the different senses described earlier, not as absolute prerequisites to reflection and discussion but as working hypotheses to see how far this can get us. As I have argued, we must also heed the imperative to "go comparative" in order to have good grounds for identifying even the basic moral modules, much less their various forms of expression, combinations, and permutations.[37] These points should lead those interested in ethics and political theory beyond the comfort of their home traditions and motivate them to open their doors and minds to respectful criticism from outside. It should also lead them to deeply and sympathetically study and critically engage other traditions and points of view and to work on developing the knowledge, skills, and virtues needed for productive engagement with alternative points of view.

Dogs and Elephants

The preceding section mentioned Jonathan Haidt's "social intuitionism" or "Moral Foundations Theory" (MFT), its relationship to the Moral Modularity Hypothesis (MMH), and the relative strength of the latter. In this section, I want to highlight and analyze the two metaphors Haidt uses to represent the relationship between reason and emotion within his theory, show how these tend to constrain a proper analysis of the issues they seek to explain, and argue that East Asian traditions offer productive ways to extend, improve, and strengthen our understanding of the complex relationship between reason and emotion. The first of Haidt's metaphors is "(the emotional) dog and its (rational) tail";[38] the second and more recent is "(the emotional) elephant and its (rational) rider."[39] The first thing to note is that there is an important ontological shift between the "dog and tail" metaphor and the "elephant and rider" metaphor. The former describes parts of a single living organism; the latter consists of two unrelated organisms and presents reason as a *foreign part* of the rider-elephant combination,[40] a part that for the most part is just along for the ride. Among other things, the former metaphor primarily conveys the relative strength between emotion and reason and seems to imply that reason is just a way to express what emotion commands; the latter metaphor strongly implies a fundamental dichotomy between emotion and reason and suggests a less integral but slightly more important role for reason, which is most unfortunate, for, as Flanagan and others have argued, Haidt does not fully appreciate the role of reason in moral life.[41] While a number of people have objected to Haidt's demotion of reason in ethical judgment, as noted earlier, both defenders and critics of Haidt's view tend to accept and perpetuate the fundamental dichotomy between reason and emotion.[42] But this is a debilitating assumption and not at all true to either psychological studies or the experiences of moral life.[43] As East Asian traditions as well as some parts of the Western tradition have argued, reason in the form of a guide, organizer, contributor, and molder of emotions—rather than some abstract, disembodied executive directing human behavior—is always and in various ways deeply involved with orienting, augmenting, extending, and shaping our emotions. This is most clearly seen in the sustained and sophisticated concern that East Asian traditions such as Confucianism have consistently shown toward ideas and practices connected with the moral cultivation of the self.[44] It is also manifest in the way that East Asian philosophers tend to talk about moral perception, judgment, and action. For example, Mengzi regularly invokes metaphors of taste to describe both the nature and path for developing moral understanding. Chapter 6 of the Confucian classic *Great Learning* (*Daxue*, 大學) describes moral insight as "like disliking a bad odor or liking something beautiful." In these examples, not only is cognition fused with emotional response, but also there is no clear moment of judging, deciding, intending, or acting. When we smell a bad odor, we immediately dislike, turn away, and distance ourselves from it.[45]

The idea that reason plays integral and critical roles as mentor, organizer, and shaper of emotions is altogether absent and not even a possibility in Haidt's first metaphor of dog and tail—a tail simply cannot play such roles in the life of a dog—and it is deeply obscured and *hidden* in his second metaphor of an elephant and its rider. The truth, however, is that the rider of Haidt's elephant enjoys the benefits of years of rational tutelage and practice, for in order for the metaphor to gain plausibility or force, it must presume that she rides a well-trained elephant; no one would ever think of, much less succeed in, riding a *wild elephant*.[46] This domesticated elephant has successfully undergone a long and complex process of training, which traditionally was done by a single person, a mahout, who typically begins working with the elephant as a young boy and almost always shares a lifelong relationship with his charge. For the most part, the mahout in turn is not relying primarily on his own reason-backed decisions. He does not devise a training program on his own. He follows a time-tested tradition of training, and here we see another way in which reason and practice are deeply embedded in the remarkable phenomenon of riding an elephant.[47] When highlighted, these points may seem obvious and become salient, but they are not at all evident in the metaphor as Haidt presents it, and here is where East Asian traditions bring something quite new and important to the table.[48]

For example, the kind of view I am advocating as an extension and enhancement of the kind of model Haidt proposes would highlight that participation in and identification with a tradition can lead us to care about being vigilant in protecting and transmitting the valuable practices and ideas that have been developed, and to feel inspired to build on and further improve them in order to prove ourselves worthy of our inheritance.[49] The care needed to protect, transmit, and further enhance what we inherit involves emotion, but ideas such as "building on" and "proving worthy of" highlight that it also involves intellectual and critical reflection. Thinking about ourselves, how we fit into our traditions, and how we can prove ourselves worthy of those in our traditions who inspire us are all aspects of the mahout module. This is a reflective, slow module—not a fast and hot one—and it is both rational and emotional.[50]

The extent to which reason and emotion are mutually implicated with one another runs deeper, and their relationship is even more complex than what has been suggested here. For example, the plausible claim that language, an ability necessary for sophisticated rational reflection, evolved as part of an ongoing effort to foster and improve cooperation and probably began with storytelling illustrates that moral emotions such as empathy, and emotions more generally, played important roles from the start in developing our most basic rational capacities.[51] Emotions of all sorts motivate, propel, and enliven such stories, and emotion and metaphor are integral features of all natural languages.[52] The work of people like Damásio has shown that reason alone cannot account for moral deliberation or judgment, so even if we hold, as we should, that reason is necessary for a genuine moral life, we must recognize not only the evolutionary connection between our more robust moral abilities and earlier more purely emotional tendencies but the ongoing roles that emotion plays in our contemporary moral deliberations.[53] What I specifically want to emphasize here is that some profoundly important roles that reason plays are embedded in humanly developed traditions and practices, and these are obscured and often overlooked in most contemporary discussions, which are cast predominantly or exclusively in terms of the *relative strength* and *contribution* of reason versus emotion.[54]

The kind of obscuration that I have argued occurs with Haidt's metaphor of an elephant and its rider also occurs in the study of East Asian moral theory, and here again a more accurate and powerful account often becomes hidden in the way certain central metaphors are deployed and understood.[55] For example, anyone familiar with East Asian Confucianism knows about Mengzi (Mencius) and his theory of human nature, which holds, among other things, that human beings are endowed with "four moral sprouts" (*siduan*, 四端) of virtue.[56] But it is critically important to keep in mind that Mengzi employs *agricultural* and not merely *vegetative* or botanical metaphors to describe the proper process of self-cultivation and that we must recognize that agriculture is based on a long and complex process of human discovery, ingenuity, and invention, which means it incorporates the accumulated insights of reason.[57] While agriculture was developed over a long period and depended on contingent environmental resources in order to take on the importance it eventually came to have, it is arguably the most important and influential invention in human history.[58] It made possible the settled, literate, large-scale community life that is most characteristic of modern human beings and posed novel and powerful demands on human behavior. When properly understood, Mengzi's metaphors, no less than Haidt's, imply that reason plays a complex and largely hidden role in our moral lives. It is hidden because the long tradition of discovery, reflection, and experimentation that constitutes the practice of agriculture is inherited by almost all human beings and is now lost in familiarity or distance from agricultural life. In regard to ethical and political theory, the norms and practices (social, political, religious, etc.) that constitute the surrounding environment in which we exist incorporate and internalize a great deal of human reasoning and give it emotional shape, direction, and power. Language itself is not just a tool for reasoning, which is how it is most often portrayed in contemporary philosophical debates. Natural human languages come preloaded with the kinds of norms, practices, and metaphors described here. These provide our starting points for moral and political reflection and theorizing, which means we all in fact begin philosophical reflection by thinking through the best views currently in play. We all start off trying to save the phenomena and should be committed to extending this exploration to the norms and practices of other successful societies and cultures. We are, or at least should be, committed to comparative philosophy.

Notes

1. I thank Karen L. Carr, Erin M. Cline, Brad Cokelet, Owen Flanagan, Eirik L. Harris, Eric L. Hutton, Bryan W. Van Norden, Hagop Sarkissian, Eric Schwitzgebel, Michael R. Slater, Justin Tiwald, and Wenqing Zhao for very helpful comments and suggestions on earlier drafts of this chapter.

2. While I use Rawls's term *reflective equilibrium* (Rawls, *The Law of Peoples*) here, I intend its sense to be much wider than his. Rawls limits the scope of reflective equilibrium to what makes sense to a particular culture or nation, perhaps worrying that this is necessary in order to avoid incoherence. Nevertheless, he presents no evidence or argument for so limiting the scope of reflective equilibrium. I see no reason to stop inquiry at the boundary of culture or tradition. To do so seems to invite not only pluralism, which I believe is both true and manageable, but strong relativism.

3. In the concluding section of this chapter, I present some reasons for thinking that the perspective of wide reflective equilibrium is an inescapable horizon for all those who philosophize about ethics and political theory using natural languages. Here, though, I am only defending the weaker claim that this in fact is the perspective from which philosophizing in these fields is practiced.

4. Haidt, The emotional dog and its rational tail; Haidt, *The Happiness Hypothesis*.

5. Haidt's results do not create problems for philosophical theory if philosophers, when they build their philosophical theories, manage to avoid the cognitive patterns he sees in nonphilosophers. Work by Eric Schwitzgebel and Fiery Cushman (Schwitzgebel and Cushman, Expertise in moral reasoning?; Schwitzgebel and Cushman, Philosophers' biased judgments persist despite training, expertise and reflection) and Josh Rust (Schwitzgebel and Rust, The moral behavior of ethicists and the power of reason) suggest that they do not manage to avoid such problems, but the issue is not yet clearly decided. The nature and extent of the problems Haidt's view would pose differ among moral theories, but if he is correct, his view would undermine the use of any theory as a reliable method for personal deliberation and choice.

6. Flanagan, *Moral Sprouts and Natural Teleologies*.

7. Engagement with East Asian traditions also lends support to certain contemporary strands of the Western philosophical tradition that have been critical of the distinction between reason and emotion or rejected it entirely. For example, Nietzsche, as well as most pragmatists, beginning with James and Dewey, rejects the distinction. I thank Michael Slater for making this point in comments on an earlier draft of this chapter.

8. The following section is taken with modification from Ivanhoe, Owen Flanagan on moral modularity and comparative philosophy.

Over time, Flanagan has made it clear that he regards theories like Kohlberg's moral psychology or the Moral Modularity Hypothesis as idealizations or models and not real descriptions of physical modules in the brain. One of his objections to them, or even to dual processing models such as Kahneman proposes (Kahneman, *Thinking Fast and Slow*), is their tendency to engender realist beliefs about analytic distinctions. This is despite the fact that Kahneman at least explicitly states that system one and system two are *not real*. I thank Owen Flanagan for comments and suggestions on this topic.

9. Chomsky, *Syntactic Structures*; Fodor, *The Modularity of Mind*.

10. Flanagan, *Varieties of Moral Personality*, 269.

11. Ibid., 270.

12. Both these features show how much Western forms of virtue ethics—even contemporary ones—still owe to Aristotle. I thank Erin M. Cline for raising this point.

13. Flanagan, *Moral Sprouts and Natural Teleologies*. Flanagan distinguishes the most general form of moral modularity, which he calls the Moral Modularity Hypothesis (MMH), from views like Haidt's, which call themselves "social intuitionism" or "Moral Foundations Theory" (MFT). He further argues that it is best to consider the former as a general hypothesis, apart from Haidt's or anyone else's specific statement of the view. Flanagan also explores with great insight and clarity the particular expression of MMH put forth by the early Confucian thinker Mengzi (391–308 BCE), which Flanagan calls Mencian Moral Modularity (MMM).

14. It is important to consider issues in childhood development here. Some capacities, for example for language and empathy, need to be engaged and developed at certain points in a child's life. There are also interesting questions in regard to developmental delays and disorders in young children, where there are normal or even exceptional capacities in some areas but serious delays or developmental problems in other areas. I thank Erin M. Cline for raising this set of issues.

15. Flanagan, *Moral Sprouts and Natural Teleologies*. This point is extremely important for those seeking to develop the MMH, for it makes clear that any adequate version of the theory requires a broad and expansive version of naturalism—one that includes the influence of nonbiological, social phenomena such as culture, history, and tradition—that might give some scientific naturalists pause. I thank Michael Slater for raising this issue.

16. Strawson, Freedom and resentment.

17. Wong, *Natural Moralities*; Ivanhoe, Pluralism, toleration, and ethical promiscuity.

18. In remarks on this section, Eric Hutton has suggested that there may be a rough parallel here with Aristotle's view about how the virtues contribute to happiness. Charitably read, Aristotle might allow that not all virtues need to be exercised in order to have a happy life; in the same way, not all moral modules need to be developed and deployed.

19. Flanagan discusses some of the ways in which modules such as disgust need to be reined in and refocused in light of what we know today (Flanagan, *Moral Sprouts and Natural Teleologies*, 62–65). Paul Bloom sees no viable moral role for disgust and argues that it often offers support for profoundly immoral behavior (Bloom, *Just Babies*, 131–157). On the other hand, the role of disgust in moral judgment has

recently been called into question. See May, Does disgust influence moral judgment?; Landy and Goodwin, Does incidental disgust amplify moral judgment?

20. Wright, *Teleological Explanations*; Lennox, Darwin and teleology.

21. Flanagan, *Moral Sprouts and Natural Teleologies*, 38–39.

22. It is common to hear people say that the theory of evolution establishes survival or the propagation of their genes as the only real value or goal of humanity, but the theory itself does no such thing. Evolutionary theory explains why some species survive and some go extinct, but it does not say that survival is good or extinction bad. I thank Eric Hutton for raising this issue.

23. Van Norden, Mengzi and Xunzi, 114.

24. Flanagan, *Moral Sprouts and Natural Teleologies*.

25. Flanagan, *The Bodhisattva's Brain*; Flanagan, Moral contagion and logical persuasion in the Mozi; Flanagan and Hu, Han Fei Zi's philosophical psychology; Flanagan and Williams, What does the Modularity of Morals have to do with ethics?; Flanagan, *Moral Sprouts and Natural Teleologies*.

26. Nisbett, *The Geography of Thought*.

27. Henrich, Heine, and Norenzayan, The weirdest people in the world?

28. Take the top twenty philosophy departments in the English-speaking world and plot the racial, gender, and ethnic diversity to be found among their senior members. Does the picture this information paints offer prima facie evidence of parochialism and perhaps a kind of guild or club of like-minded fellows or evidence of a community of objective, open-minded, clear-thinking inquirers, selected in a fair and inclusive manner?

29. Flanagan, *Moral Sprouts and Natural Teleologies*, 76.

30. Eric Hutton (Moral connoisseurship in Mengzi) adds the important point that in some cases ancient observers saw things *more* clearly than many people nowadays do. We can, and perhaps tend to, be so absorbed in technology and the problems it creates that we overlook the ways in which the ancients, because of the simpler lives they led, had greater insight into widespread and fundamental human problems.

31. For an insightful study of Leibniz's views about the importance of China for philosophy, see Perkins, *Leibniz and China*.

32. For a thorough study that reveals the shallow and despicable reasons that motivated the historical neglect of non-Western philosophy, see Park, *Africa, Asia, and the History of Philosophy*.

33. I am indebted to Eirik L. Harris for raising the issues explored in the following paragraph.

34. Richard Rorty explicitly defended something like this view. See, for example, Rorty, *Contingency, Irony, and Solidarity*.

35. Flanagan considers a possible actual example of this kind of culture in Flanagan, *Moral Sprouts and Natural Teleologies*.

36. Ivanhoe, Pluralism, toleration, and ethical promiscuity.

37. It is an open question what we might learn from the comparative program. Minimally, we will learn different ways and styles of expressing and combining whatever moral modules we might possess, but we might also learn things about the nature, type, and range of the basic models themselves. An example of how comparative study can directly contribute to our understanding of even basic human capacities is the relatively recent addition of umami ("savory taste") as one of the five basic tastes (along with sweetness, sourness, bitterness, and saltiness). Umami is the romanization for the Japanese term うま味, which means "pleasant savory taste." It was coined by Kikunae Ikeda, who first proposed the existence of umami in 1908. The fact that leading scholars such as Haidt continue to add to and adjust their accounts of the moral modules shows that it would be premature to cut off the potential value of comparative study. I thank Eric L. Hutton and Justin Tiwald for discussions and suggestions on this issue.

38. Haidt, The emotional dog and its rational tail.

39. Haidt, *The Happiness Hypothesis*. Haidt notes the earlier precedent of Plato's metaphor in the *Phaedrus* of the self or soul as a chariot, with reason as a chariot driver holding the reins and trying to control and coordinate two horses (spirit and appetite) that do the pulling; he has very interesting and insightful things to say about this and other models of the self (*The Happiness Hypothesis*, 2–5). Haidt also makes it clear that he chose the elephant and rider to highlight how little control reason has when emotion or desire is in play (ibid., 4).

40. Eric Hutton has reminded me that for Plato there is a strong sense in which reason really is a distinct and *foreign* part of the human composite, for Plato thinks of reason as something that is *divine* and superior to the rest of our being.

41. Another way of thinking about the difference between the metaphors is to say that in the case of the dog and its tail, practical reasoning is just an indicator or expression of the emotions, while in the case of the elephant and rider, practical reasoning sets the goals and determines the means for achieving them, while the elephant provides the motivation to get there. The latter is closer to but distinct from Hume's notion that practical reason is a slave to the passions but can determine the means for achieving goals and identify possible goals to pursue. I thank Bryan W. Van Norden for suggesting this way of unpacking Haidt's metaphors. Haidt's model does allow for one's reasoning to affect *other people's* emotions and intuitions—young and impressionable people who observe or listen to that person—so at least

in this respect he does allow for contributions by reason. In a personal conversation with me, Eric Schwitzgebel noted that Haidt raised this point.

42. This is true of just about every contemporary treatment of this theme, even such rich and revealing works as de Waal, *Good Natured*; de Waal, *Primates and Philosopher.*

43. In more recent work (Haidt, *The Righteous Mind*), Haidt has begun referring to the "intuitive" elephant rather than the "emotional" elephant. This is an important change, which seems to reflect his stepping away from the emotion/reason dichotomy and toward something more like the fast/slow dichotomy first described by Kahneman. This suggests that Haidt might already be moving partly in the direction advocated in this chapter. I thank Eric Schwitzgebel for this insight.

44. Ivanhoe, *Confucian Moral Self Cultivation.*

45. This kind of merging of cognition, feeling, judging, deciding, intending, and acting into a single seamless process was actively discussed throughout the tradition. One of the most dramatic positions on this set of issues was taken by Wang Yangming (1472–1529), who advocated "the unity of knowing and acting." For a description and analysis of his view, see Ivanhoe, Moral perception in McDowell, Wang, and Mengzi.

46. The kind of view I advocate is at odds with the picture of emotions as part of a "fast" or "automatic" system if this is understood as largely freestanding, isolated, and fixed. Our affective system is more flexible, broad, and can develop over time. For a defense of such a view and its implications, see Railton, The affective dog and its rational tale. See also Flanagan, *Moral Sprouts and Natural Teleologies.*

47. Of course, the mahout does rely on practical reason in applying traditional methods of training and may engage in theoretical reflection on how to improve what he has inherited, but the point remains. We might consider whether our contemporary psychological model should include a "mahout module"—roughly a reflective self-guidance module. Such a module would not initially have a specified content but would describe a capacity or set of capacities that are the result of evolutionary forces. This module may itself turn out to have various aspects and be understood differently in different traditions, but all empirically informed philosophers should grant that this mahout, when fully mature or developed, is something that is the result of habituation, learning, reflection, and discipline. I thank Brad Cokelet for this suggestion.

48. Kwongloi Shun notes that there is asymmetry between the rate at which ideas derived from East Asian philosophy are applied to the study of Western philosophy and the rate at which ideas derived from Western philosophy are applied to the study of East Asian philosophy (Shun, Studying Confucian and comparative ethics). I agree with this important general point, though note that there are exceptions. These include work applying East Asian conceptions of moral cultivation of the self (Ivanhoe, *Confucian Moral Self Cultivation*; Cline, *Families of Virtue*), spontaneity

(Ivanhoe, The values of spontaneity), connoisseurship (Hutton, Moral connoisseurship in Mengzi), ritual (Sarkissian, Minor tweaks, major payoffs; Tiwald and Kline, *Ritual and Religion in the Xunzi*), and political theory (Cline, *Confucius, Rawls, and the Sense of Justice*; Chan, *Confucian Perfectionism*). For a general discussion of the nature, challenges, and potential rewards of comparative philosophy, see Van Norden, *Virtue Ethics and Consequentialism in Early Chinese Philosophy*; Cline, *Confucius, Rawls, and the Sense of Justice* ; Wong, Comparative philosophy.

49. I advance such an argument, and offer several examples of cases where such concerns are in play, in Ivanhoe, *Confucian Reflections*, 1–16.

50. I thank Brad Cokelet for suggesting this line of thought and especially for pointing out the relationship between this and my earlier writings on tradition.

51. Gibbard, *Wise Choices, Apt Feelings*.

52. Lawrence, The sacred bee, the filthy pig, and the bat out of hell.

53. Damásio, *Descartes' Error*.

54. The position I am advocating is deeply opposed to the commonly encountered view that traditions are largely *impediments* to reason, at least in its "proper" critical employment. Rather, the idea is that traditions can actually exhibit rationality or generally function as reliable guides to good reasoning. They can perform the role of cognitively "off-loading" a good deal of work. My view is related to but distinct from Alasdair MacIntyre's idea (MacIntyre, *Whose Justice? Which Rationality?*) that different traditions embody distinct forms of rationality. I thank Michael Slater for helping me to see and express these ideas.

55. Unpacking and carefully analyzing these and other metaphors at the heart of different philosophical or psychological views is an important task for, as David B. Wong notes, "metaphors do not confirm themselves" (Wong, Early Confucian philosophy and the development of compassion, 191). Several of my works on East Asian and comparative philosophy (Ivanhoe, *Ethics in the Confucian Tradition*; Ivanhoe, *Confucian Moral Self Cultivation*) involve the careful unpacking and analysis of what I take to be some of the guiding metaphors in the Confucian tradition. Others have produced similar works, though, with different assumptions about the nature of metaphors. For example, some seem to believe that the use of metaphors is distinctive and perhaps even unique to Chinese philosophy (Allan, *The Way of Water and Sprouts of Virtue*), while others employ views such as Conceptual Metaphor Theory and argue that certain core metaphors are shared by all human beings (Slingerland, Metaphor and meaning in early China). My approach lies between these alternatives; I believe metaphors often play critical roles in philosophical theories and are part of every human language, but I hold no theory about deep, shared, universal metaphors informing human cognition. For an example of the kind of approach I employ, see Munro, *Images of Human Nature*.

56. There are important related points involving the proper translation of this term, which I have explored in other work (Ivanhoe, *Confucian Moral Self Cultivation*, 18–20), but here I want to focus on how this metaphor is deployed and understood.

57. Ivanhoe, *Confucian Reflections*. Flanagan (Flanagan, *Moral Sprouts and Natural Teleologies*, 59–62) recently suggested a number of interesting ways to extend and enrich Mengzi's agricultural metaphors.

58. For a most insightful study of the importance of agriculture and its course of development, see Diamond, *Guns, Germs, and Steel.*

Bibliography

Allan, Sarah. *The Way of Water and Sprouts of Virtue*. Albany: State University of New York Press, 1997.

Bloom, Paul. *Just Babies: The Origins of Good and Evil*. New York: Crown, 2014.

Chan, Joseph. 2014. *Confucian Perfectionism: A Political Philosophy for Modern Times*. Princeton, NJ: Princeton University Press, 2014.

Chomsky, Noam. *Syntactic Structures*. The Hague: Mouton, 1957.

Cline, Erin M. *Confucius, Rawls, and the Sense of Justice*. New York: Fordham University Press, 2013.

Cline, Erin M. *Families of Virtue: Confucian and Western Views on Childhood Development*. New York: Columbia University Press, 2015.

Damásio, António. *Descartes' Error: Emotion, Reason, and the Human Brain*. 1994. Reprint. New York: Penguin Books, 2005.

de Waal, Frans B. M. *Good Natured: The Origins of Right and Wrong in Humans and Other Animals*. Cambridge, MA: Harvard University Press, 1996.

de Waal, Frans B. M. *Primates and Philosophers: How Morality Evolved*. Edited by Stephen Macedo and Josiah Ober. Princeton, NJ: Princeton University Press, 2006.

Diamond, Jarrad M. *Guns, Germs, and Steel: The Fates of Human Societies*. London: Jonathan Cape, 1997.

Flanagan, Owen. *The Bodhisattva's Brain: Buddhism Naturalized*. Cambridge, MA: MIT Press, 2011.

Flanagan, Owen. Moral contagion and logical persuasion in the Mozi. *Journal of Chinese Philosophy* 35, no. 3 (2008): 473–491.

Flanagan, Owen. *Moral Sprouts and Natural Teleologies: 21st Century Moral Psychology Meets Classical Chinese Philosophy* (The Aquinas Lecture, 2014). Milwaukee, WI: Marquette University Press, 2014.

Flanagan, Owen. *Varieties of Moral Personality: Ethics and Psychological Realism*. Cambridge, MA: Harvard University Press, 1991.

Flanagan, Owen, and Hu, J. Han Fei Zi's philosophical psychology: Human nature, scarcity, and the neo-Darwinian consensus. *Journal of Chinese Philosophy* 38, no. 2 (2011): 293–316.

Flanagan, Owen, and Williams, R. A. What does the Modularity of Morals have to do with ethics? Four moral sprouts plus or minus a few. *Topics in Cognitive Science* 2, no. 3 (2010): 430–453.

Fodor, J. A. *The Modularity of Mind*. Cambridge, MA: MIT Press, 1983.

Gibbard, Allan. *Wise Choices, Apt Feelings*. Cambridge, MA: Harvard University Press, 1990.

Haidt, Jonathan. The emotional dog and its rational tail: A social intuitionist approach to moral judgment. *Psychological Review* 108, no. 4 (2001): 814–834.

Haidt, Jonathan. *The Happiness Hypothesis: Finding Modern Truth in Ancient Wisdom*. New York: Basic Books, 2006.

Haidt, Jonathan. *The Righteous Mind: Why Good People Are Divided by Politics and Religion*. New York: Pantheon, 2012.

Henrich, J., Heine, S. J., and Norenzayan, A. The weirdest people in the world? *Behavioral and Brain Sciences* 33, no. 2/3 (2010): 61–83.

Hutton, Eric L. Moral connoisseurship in Mengzi. In *Essays on the Moral Philosophy of Mengzi*, edited by Xiusheng Liu and Philip J. Ivanhoe, 163–186. Indianapolis, IN: Hackett, 2002.

Ivanhoe, Philip J. *Confucian Moral Self Cultivation*. Rev. 2nd ed. Indianapolis, IN: Hackett, 2006.

Ivanhoe, Philip J. *Confucian Reflections: Ancient Wisdom for Modern Times*. London: Routledge and Kegan Paul, 2013.

Ivanhoe, Philip J. *Ethics in the Confucian Tradition: The Thought of Mengzi and Wang Yangming*. Rev. and exp. 2nd ed. Indianapolis, IN: Hackett, 2002.

Ivanhoe, Philip J. Moral perception in McDowell, Wang, and Mengzi. *Dao: A Journal of Comparative Philosophy* 10, no. 3 (2011): 273–290.

Ivanhoe, Philip J. Owen Flanagan on moral modularity and comparative philosophy. In *Naturalism, Human Flourishing, and Asian Philosophy: Owen Flanagan and Beyond*, edited by Bongrae Seok, 208–223. New York: Routledge, 2019.

Ivanhoe, Philip J. Pluralism, toleration, and ethical promiscuity. *Journal of Religious Ethics* 37, no. 2 (2009): 311–329.

Ivanhoe, Philip J. The values of spontaneity. In *Taking Confucian Ethics Seriously: Contemporary Theories and Applications*, edited by Yu Kam-por, Julia Tao, and Philip J. Ivanhoe, 183–207. Albany: State University of New York Press, 2010.

Kahneman, Daniel. *Thinking Fast and Slow*. New York: Farrar, Straus and Giroux, 2011.

Landy, J. F., and Goodwin, G. P. Does incidental disgust amplify moral judgment? A meta-analytic review of experimental evidence. *Perspectives on Psychological Science* 10, no. 4 (2015): 518–536.

Lawrence, Elizabeth Atwood. The sacred bee, the filthy pig, and the bat out of hell: Animal symbolism as cognitive biophilia. In *The Biophilia Hypothesis*, edited by Stephen R. Kellert and Edward O. Wilson, 301–344. Washington, DC: Island Press, 1993.

Lennox, James G. Darwin and teleology. In *The Cambridge Encyclopedia of Darwin and Evolutionary Thought*, edited by Michael Ruse, 152–157. Cambridge: Cambridge University Press, 2013.

MacIntyre, Alasdair C. *Whose Justice? Which Rationality?* Notre Dame, IN: University of Notre Dame Press, 1988.

May, Josh. Does disgust influence moral judgment? *Australasian Journal of Philosophy* 92, no. 1 (2014): 125–141.

Munro, Donald J. *Images of Human Nature: A Sung Portrait*. Princeton, NJ: Princeton University Press, 1988.

Murdoch, Iris. *The Sovereignty of Good*. 1970. Reprint. London: Routledge, 1991.

Nisbett, Richard. *The Geography of Thought: How Asians and Westerners Think Differently … and Why*. New York: Free Press, 2003.

Park, Peter K. J. *Africa, Asia, and the History of Philosophy: Racism in the Formation of the Philosophical Canon, 1780–1830*. Albany: State University of New York Press, 2013.

Perkins, Franklin. *Leibniz and China: A Commerce of Light*. New York: Cambridge University Press, 2004.

Railton, Peter. The affective dog and its rational tale: Intuition and attunement. *Ethics* 124, no. 4 (July 2014): 813–859.

Rawls, John. *The Law of Peoples*. Cambridge, MA: Harvard University Press, 1999.

Rorty, Richard. *Contingency, Irony, and Solidarity*. Cambridge, MA: Cambridge University Press, 1989.

Sarkissian, Hagop. Minor tweaks, major payoffs: The problems and promise of situationism in moral philosophy. *Philosopher's Imprint* 10, no. 9 (2010): 1–15.

Schwitzgebel, Eric, and Cushman, Fiery. Expertise in moral reasoning? Order effects on moral judgment in professional philosophers and non-philosophers. *Mind and Language* 27, no. 2 (April 2012): 135–153.

Schwitzgebel, Eric, and Cushman, Fiery. Philosophers' biased judgments persist despite training, expertise and reflection. *Cognition* 141 (August 2015): 127–137.

Schwitzgebel, Eric, and Rust, Joshua. The moral behavior of ethicists and the power of reason. In *Advances in Experimental Moral Psychology*, edited by Hagop Sarkissian and Jennifer Cole Wright, 91–109. New York: Bloomsbury, 2014.

Shun, Kwong-loi. Studying Confucian and comparative ethics: Methodological reflections. *Journal of Chinese Philosophy* 36, no. 3 (2009): 455–478.

Slingerland, Edward. Metaphor and meaning in early China. *Dao: A Journal of Comparative Philosophy* 10, no. 1 (2011): 1–30.

Strawson, P. F. Freedom and resentment. *Proceedings of the British Academy* 48, no. 1 (1962): 1–25.

Tiwald, Justin, and Kline, Thorton C., III, eds. *Ritual and Religion in the Xunzi*. Albany: State University of New York Press, 2014.

Van Norden, Bryan W. Mengzi and Xunzi: Two views of human agency. In *Virtue, Nature, and Moral Agency in the Xunzi*, edited by T. C. Kline III and Philip J. Ivanhoe, 103–134. Indianapolis, IN: Hackett, 2000.

Van Norden, Bryan W. *Virtue Ethics and Consequentialism in Early Chinese Philosophy*. Cambridge: Cambridge University Press, 2007.

Wong, D. B. Comparative philosophy: Chinese and Western. In *The Stanford Encyclopedia of Philosophy*, edited by Edward N. Zalta. Stanford, CA: Stanford University, 2014. http://plato.stanford.edu/archives/win2014/entries/comparphil-chiwes/.

Wong, D. B. Early Confucian philosophy and the development of compassion. *Dao: A Journal of Comparative Philosophy* 14, no. 2 (2015): 157–194.

Wong, D. B. *Natural Moralities: A Defense of Pluralistic Relativism*. Princeton, NJ: Princeton University Press, 2006.

Wright, L. *Teleological Explanations*. Berkeley: University of California Press, 1976.

Contributors

Jack Bauer, professor of psychology, University of Dayton

Patricia S. Churchland, professor emerita of philosophy, University of California at San Diego, and adjunct professor, Salk Institute

Peggy DesAutels, professor of philosophy, University of Dayton

George Graham, former professor of philosophy, University of Alabama, Birmingham; Wake Forest University; and Georgia State University

Philip J. Ivanhoe, Distinguished Chair Professor, College of Confucian Studies and Eastern Philosophy, and director of the Sungkyun Institute for Confucian Studies and East Asian Philosophy

Alasdair MacIntyre, professor emeritus, Duke University, and Rev. John A. O'Brien Senior Research Professor of Philosophy emeritus, University of Notre Dame

Robert N. McCauley, William Rand Kenan Jr. University Professor, Emory University

Eddy Nahmias, professor of philosophy, Georgia State University

Thomas W. Polger, professor of philosophy, University of Cincinnati

Galen Strawson, professor of philosophy, University of Texas, Austin

Şerife Tekin, assistant professor of philosophy, University of Texas, San Antonio

Robert Van Gulick, professor of philosophy, Syracuse University

David B. Wong, Susan Fox Beischer & George D. Beischer Professor of Philosophy, Duke University

Wenqing Zhao, assistant professor of philosophy, Whitman College

Index